# Living the Ancient Southwest

# Living the Ancient Southwest

*Edited by David Grant Noble*

A School for Advanced Research Popular Archaeology Book

School for Advanced Research Press
Santa Fe

School for Advanced Research Press
Post Office Box 2188
Santa Fe, New Mexico 87504-2188
www.sarpress.org

Series Editor: Lisa Pacheco
Managing Editor: Lisa Pacheco
Editorial Assistant: Ellen Goldberg
Designer and Production Manager: Cynthia Dyer
Proofreader: Kate Whelan
Indexer: Ina Gravitz
Printed by Sheridan Books

Library of Congress Cataloging-in-Publication Data
Living the ancient Southwest / edited by David Grant Noble. — First edition.
pages cm
Summary: "How did Southwestern peoples subsist in the arid reaches of the Great Basin? When and why did violence erupt in the Mesa Verde region? Who were the Fremont people? How do some Hopis view Chaco Canyon? These are a few of the topics addressed in Living the Ancient Southwest. The essayists in this new highly-illustrated anthology also write about the beauty and originality of Mimbres pottery, the rock art in Canyon de Chelly, the history of the Wupatki Navajos, and O'odham songs describing ancient trails to the Pacific Coast. The anthropologist-writers in this book will enlighten readers on a variety of subjects relating to the deep history and culture of the American Southwest"—Provided by publisher.
"A School for Advanced Research popular archaeology book."
Includes index.
ISBN 978-1-938645-45-7 (cloth : alkaline paper) — ISBN 978-1-938645-46-4 (paper : alkaline paper)
1. Indians of North America—Southwest, New—Antiquities. 2. Indians of North America—Southwest, New—History. 3. Indians of North America—Southwest, New—Social life and customs. 4. Southwest, New—Antiquities. 5. Southwest, New—Social life and customs. 6. Southwest, New—Environmental conditions. 7. Human ecology—Southwest, New—History. 8. Social archaeology—Southwest, New. 9. Ethnoarchaeology—Southwest, New. I. Noble, David Grant.
E78.S7L635 2014
979'.01—dc23
2014025154

© 2014 by the School for Advanced Research

All rights reserved.
Manufactured in the United States of America
International Standard Book Numbers
cloth 978-1-938645-45-7
paper 978-1-938645-46-4
First edition 2014  Second paperback printing 2019

Cover photograph: Aerial of Chimney Rock Pueblo © Adriel Heisey. Used with permission.

# Contents

| | | |
|---|---|---|
| | Preface<br>David Grant Noble | vii |
| | Acknowledgments<br>David Grant Noble | viii |
| | Map 1. Ancient Southwest Sites | ix |
| *one* | Making a Living in the Desert West<br>Steven R. Simms | 1 |
| *two* | Pueblo Farmers of the Chacoan World<br>R. Gwinn Vivian | 11 |
| *three* | Through the Looking Glass: The Environment of the Ancient Mesa Verdeans<br>Karen R. Adams | 19 |
| *four* | Ancient Violence in the Mesa Verde Region<br>Kristin A. Kuckelman | 27 |
| *five* | Carved in the Cliffs: The Cavate Pueblos of Frijoles Canyon<br>Angelyn Bass | 37 |
| *six* | Architecture: The Central Matter of Chaco Canyon<br>Stephen H. Lekson | 45 |
| *seven* | Wupatki Pueblo: Red House in Black Sand<br>Christian E. Downum, Ellen Brennan, and James P. Holmlund | 55 |
| *eight* | Zuni Religion and Philosophy<br>Edmund J. Ladd | 65 |
| *nine* | Yupköyvi: The Hopi Story of Chaco Canyon<br>Leigh J. Kuwanwisiwma | 73 |

| | | |
|---|---|---|
| *ten* | Canyon de Chelly: A Navajo View<br>    An Interview of Mrs. Mae Thompson by Irene Silentman | 81 |
| *eleven* | The Wupatki Navajos: An Historical Sketch<br>    Alexa Roberts | 89 |
| *twelve* | The Enigmatic Fremont<br>    Joel C. Janetski | 97 |
| *thirteen* | The Hohokam Millennium<br>    Suzanne K. Fish and Paul R. Fish | 107 |
| *fourteen* | Pottery of the Sierra Sin Agua<br>    Kelley Hays-Gilpin and Christian E. Downum | 119 |
| *fifteen* | Expressions in Black and White<br>    Michelle Hegmon | 127 |
| *sixteen* | Ancestral Pueblo Rock Art in Tsegi Canyon and Canyon de Chelly:<br>A View behind the Image<br>    Polly Schaafsma | 137 |
| *seventeen* | Songscapes and Calendar Sticks<br>    J. Andrew Darling and Barnaby V. Lewis | 149 |
| *eighteen* | Ancient Trails of the Pajarito Plateau<br>    James E. Snead | 159 |
| | Picture Credits | 166 |
| | Chapter Credits | 168 |
| | Index | 169 |

Color plates follow page 54.

# Preface
## David Grant Noble

More than forty years ago, a door opened for me and I found myself, quite suddenly, in the field of archaeology. During several summers, my job was to document with camera and film the excavations at Arroyo Hondo Pueblo near Santa Fe, New Mexico. While young suntanned archaeologists carefully picked and scraped at room profiles, uncovered long-dormant hearths, and slung dirt in wheelbarrows, I was on the spot taking pictures. There were many artifacts to be recorded, as well: pots and grinding stones, bone awls and shell ornaments, wood beams and burials. From time to time, excitement ran through the crew as something special—a macaw skeleton, a turkey nest containing eggs, a room stacked with burnt corncobs—was unearthed. Then, for *National Geographic Magazine*, pictures were needed of camp life, scenery, and archaeologists at play. Having such experiences, how could I resist catching some form of the archaeological bug?

Later, when I joined the staff of the School of American Research (now the School for Advanced Research), I was editor of its annual membership bulletin, *Exploration*. In 1980, we began focusing each issue on a particular Southwestern archaeological preserve and persuaded scholars to contribute essays addressed to general readers with an avid interest in Southwestern history and culture. These modest publications eventually evolved into books, such as *New Light on Chaco Canyon* and *The Hohokam: Ancient People of the Desert*.

In 2002, at the Pecos Conference, I met James F. Brooks, then director of SAR Press. Our conversation gave rise to the Press's Popular Archaeology Series, which has produced seven volumes to date including 124 essays. In these books, many leading scholars and other specialists share their knowledge of the American Southwest with a broad and diverse readership. Together with the *Exploration* articles, their richly illustrated chapters covering a wide variety of topics have helped to expand public understanding and appreciation of the Southwest's deep cultural heritage. They have also provided the wellspring for the book you now hold in your hands.

For me, the process of reviewing so much scholarship was like taking an advanced degree in anthropology, and the experience reinvigorated my long-held enthusiasm for the subject. My goal was to bring into being a book that would introduce readers to the Southwest by providing insights into the region's Native peoples and showing some of the beautiful architecture, ceramics, rock art, and implements they created.

Some essays in the following pages address this artistic creativity: the beauty and significance of Southwestern peoples' minds and hands at work. Others discuss how people met the challenges of subsisting in arid and seemingly desolate environments. Several convey Native American viewpoints on the history and significance of places on the landscape, and one author offers insights into the religion and philosophy of his own tribe. Two unusual chapters discuss ancient trail networks, two focus on the less well-known Fremont and Hohokam cultures, and still another treats the delicate topic of violence and warfare.

I have lived in the Southwest for more than forty years and love its beautiful landscapes and rich cultural heritage. I wish for this book to help us all cherish and protect it and to honor its diverse peoples.

# Acknowledgments
## *David Grant Noble*

First and foremost, I thank the authors herein for sharing their research and knowledge in a way that general readers will appreciate. The volume editors of the books in SAR Press's Popular Archaeology Series deserve special recognition for bringing together the volumes in this valuable and well-received series.

Generously, many individuals granted permission to publish gratis their photographs and illustrations. The School for Advanced Research and I extend our thanks to Steven R. Simms, R. Gwinn Vivian, Bruce Hucko, Patricia McCreery, Laurie Logsdon, Peter J. Pilles Jr., Irene Silentman, Alexa Roberts, Polly Schaafsma, Matts Myhrman, Christian E. Downum, Will Russell, Russ Bodner, Carolyn Kernberger, J. Andrew Darling, Barnaby V. Lewis, and James Snead.

We also thank the following institutions for providing illustrations gratis: Archives, Museum of Indian Arts and Culture, Museum of New Mexico; National Park Service; Museum of Northern Arizona; American Museum of Natural History; Crow Canyon Archaeological Center; University of Utah Marriott Library Special Collections; Museum of Peoples and Cultures, Brigham Young University; Arizona State Museum; Cline Library, Northern Arizona University; Bilby Research Center, Northern Arizona University; Amerind Foundation, Inc.; Western New Mexico University Museum; Cultural Resource Management Program, Gila River Indian Community; National Museum of the American Indian; and Denver Public Library, Western History Collection.

It has been a pleasure to work collaboratively with SAR Press staff members Lisa Pacheco, Ellen Goldberg, Cynthia Dyer, and Lynn Baca. SAR librarian Laura Holt was helpful and deserves credit for cataloging and conserving photographs used to illustrate issues of *Exploration* published more than twenty years ago. I thank Jennifer Day at the Indian Arts Research Center for her assistance in providing illustrations. I also wish to recognize former SAR president James F. Brooks, who initiated and developed the Popular Archaeology Series while he was director of the Press and provided editorial direction and valuable insights as the series progressed.

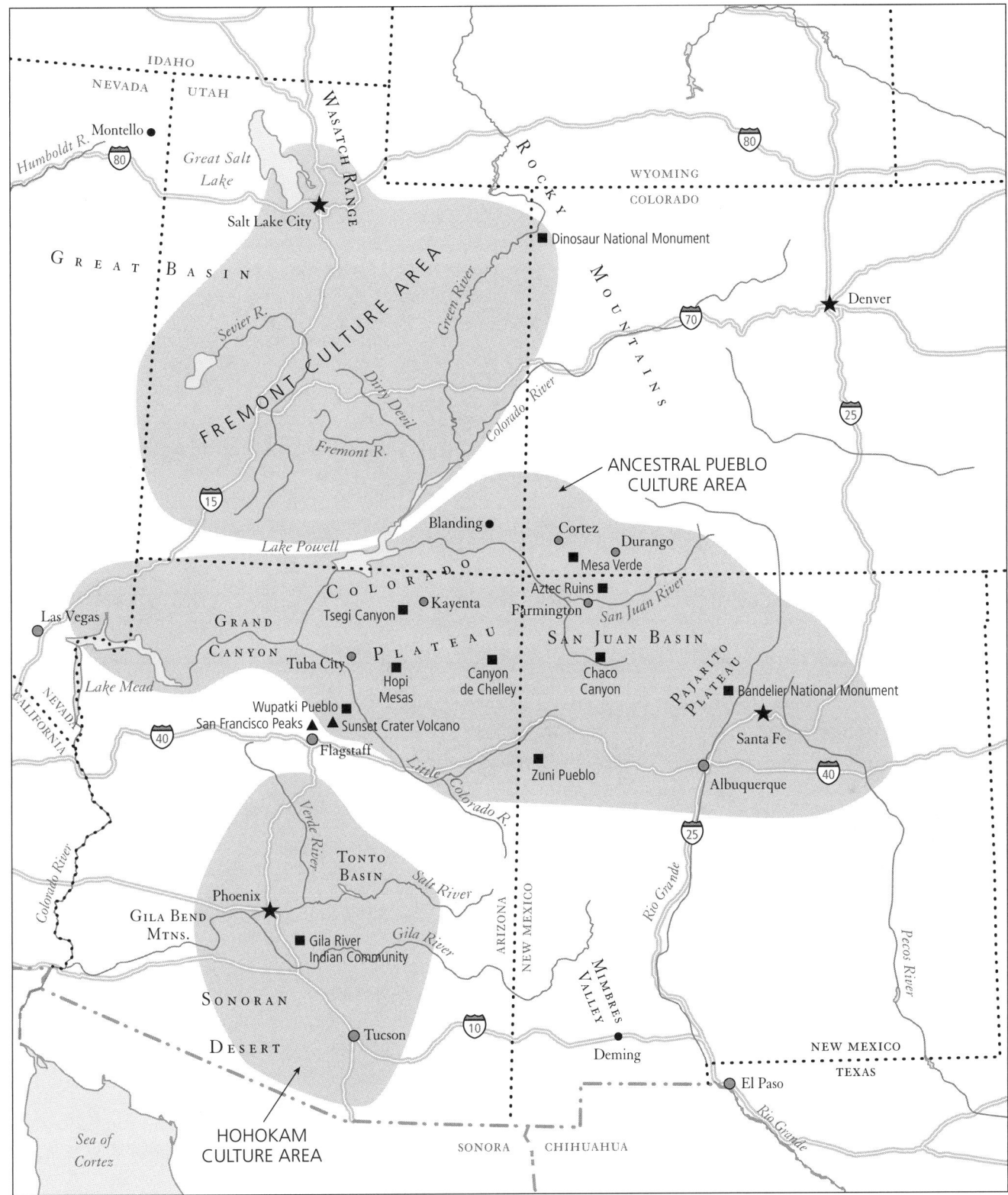

Map 1. Ancient Southwest Sites.

Figure 1.1. Pronghorn antelope trap near Montello, Nevada. The juniper branches are remnants of the trap.

# Making a Living in the Desert West

*Steven R. Simms*

Our fifteen-passenger van rolled to a stop on a small rise overlooking a sage-covered plain. This place looked like all the others we had jostled past for the last hour on the seemingly endless dirt roads of northeastern Nevada. We were part of a university course titled "Archaeology and Paleoenvironments Field Trip," but this was no arrowhead hunting trip. We wanted to see how people lived in the ancient Great Basin. Everyone craned their necks and their eyes searched when I declared, "We're here! Can you see it?" The students tumbled out of the van as I began walking a line of broken juniper branches made gray by three centuries of cold, dry desert air. The branches lay jumbled upon one another, but together they marched through the sagebrush in a long, gently curving line.

The branches were the remnants of a toppled fence—a pronghorn trap, a place where Shoshones hunted the misnamed antelope by driving them into large enclosures. We walked a third of a mile along the discontinuous alignments of branches forming the length of the trap and traced the quarter mile of its width. From the air, the trap looks like a giant keyhole. It opens below the small knoll where we parked, and the collapsed fence curves around the base of the knoll to create a natural funnel that surely once steered animals to their death.

This pronghorn trap is about three hundred years old. Like most others, it takes advantage of the terrain and the natural behavior of the animals, demonstrating that hunters of the past knew their prey well. Pronghorn typically escape their predators by following the lowest ground, sometimes even appearing to skulk a bit to be less visible. This trap was built around a subtle swale at the bottom of a long, shallow slope down a valley. Hunters drove the animals here from miles away, probably steering them with piles of brush festooned with strips of bark that fluttered in the breeze and an occasional human "beater" to keep them on the right path.

The hunt likely started the day before the actual kill, with people moving the animals naturally but deliberately. A pronghorn headdress worn by a shaman helped charm the animals into proximity by playing on their natural curiosity. As the pronghorn neared the trap, the hunters closed in, causing their prey to accelerate into their characteristic high-speed run. The animals hugged the swale and unknowingly entered the enclosure, only to be surprised by people hiding around its perimeter. The fence of branches was only waist high, but the

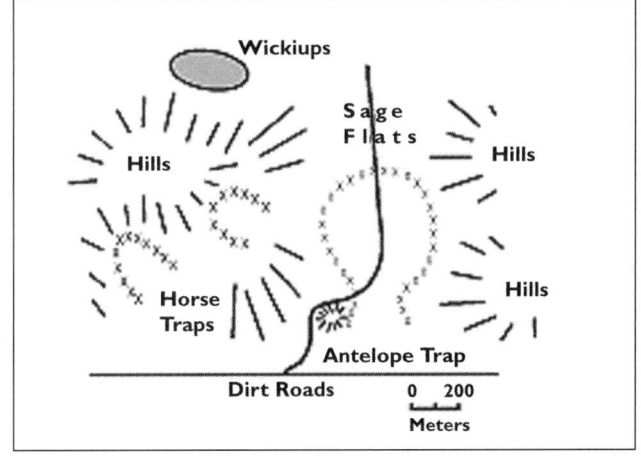

Figure 1.2. Sketch of the antelope trap north of Montello, Nevada.

Figure 1.3. A wickiup, or log and brush shelter, at the Bustos site near Ely, Nevada. This was one of five wickiups at the site, a pine nut–gathering camp used in the fall. Eight circular, rock-lined storage bins were found nearby.

builders knew that instead of leaping the fence, the pronghorn would obligingly turn in order to maintain full tilt. (This response to barriers is one reason pronghorn suffered so much as fences parceled out the American West for cattle.) Once hunters surrounded the animals, they forced them into a small herd on the lowest ground near the middle of the corral. Hunters took several dozen in a hunt like this, picking them off one by one with arrows or forcing them to run until they collapsed from exhaustion and could be clubbed.

The remnants of a small village of log and brush houses, or wickiups, stand about a quarter mile from the pronghorn trap. Easily refurbished whenever hunters wanted to use the trap, the village served as home base during the hunt. For days or even weeks afterward, people remained in the village to process the animals for dried meat, hides, and sinews. They boiled the bones for grease or fashioned them into awls and fishhooks or pendants and beads. When people departed, they left behind some of the things they had needed for the hunt. They always carried stone from the best quarries, to be flaked into tool blanks and perhaps cached at a village for the next hunting trip. People also cached fiber for ropes and string, extra arrows and tips, and perhaps even snares that could be set and left to work while people hunted the pronghorn.

People used the village to prepare for the hunt and repair their equipment afterward. One such place near a pronghorn trap in eastern California yielded the broken base of an arrowhead that fit perfectly with a tip archaeologists found inside the trap. The tip must have lodged inside a slaughtered animal, and when the hunter returned to camp, the broken base was removed from the arrow shaft and discarded.

Several years or even a decade might have passed before hunters used this trap again, but it was only one of many traps placed strategically across the landscape. Archaeologists have records of dozens of pronghorn traps in northeastern Nevada alone. They were common features of the "built environment" of the indigenous peoples of the Great Basin.

## A Full and Rich Life

The Great Basin is a land so harsh that many modern people passing through it seem reluctant to get out of their cars. US Highway 50 across central Nevada is billed as "the loneliest road in America" on bumper stickers sold at gas stations and rock shops. In our modern world of comfort and insulation from nature, which extends even to the outdoorsy among us, it is difficult to imagine people living in such a place, let alone living well. But the ancients were no mere survivalists wandering desperately in search of food. Their thirteen thousand-year history, spanning more than four hundred human generations, testifies to their success at living in the Great Basin. The reason they thrived there is that they knew the region intimately. It was a human landscape.

The people had a geographically expansive sense of place. They envisioned the land in terms of homelands rather than of single, fixed homesites. The landscape was socially full, enveloped by a network that cycled people among kin and place. The sizes of their groups pulsed according to circumstances, and broad notions of kinship ensured connections even in a land with some of the lowest population densities ever recorded.

People were nomadic, but their movements were not aimless. Where they lived and moved was structured by the seasonal scheduling of activities, social expectations, religious rules, and an intimate knowledge of nature, all of which formed a seamless whole. For instance, the sequence of ripening plants had to be followed, but an opportunity to harvest a sagebrush flat rich in cottontail rabbits would not be passed up. Much of what people did in the summer was in anticipation of winter, when stored food supplies were essential. They lived in some places for months and in others for only weeks or days. Such decisions depended as much on kin and other social dynamics as they did on where the food was—but then those things, too, were inextricably connected.

Ancient people did not have the same notions of private property as modern Americans, and they defined territoriality largely by use, not by permanent residence. Most resources and property were public goods that people shared. During their lifetime, people might weave many territories into their lived experiences. Relations of kinship and social obligations hovered over the land like a net. Where fences now divide the land into private parcels, the landscape then was a stage on which interactions of cooperation, but also competition and even conflict, shaped who lived where and who decided what would happen. A built environment specialized to fit many places and uses complemented this pliable social fabric and means of making a living.

## The Built Environment

The built environment of our modern world is so much a part of life that we give it scarcely any thought. Imagine life without our homes, roads, churches, baseball fields, and schools. Where would we be without sewers, fiber optic cables, electrical generating plants, and factories? We rely on our infrastructure.

The early peoples of the Great Basin also had built environments. They did not have to carry all their worldly possessions on their backs. Instead, they cached gear and supplies in places they knew they would return to. When it was time to hunt in the marshes, they could swing by a cache containing net bags and snares to capture small animals, as well as fishing lines, hooks, and weights. They might have left in storage bone tubes for snorkels and duck decoys like those woven from cattail stalks two thousand years ago and left in Lovelock Cave, Nevada (plate 2).

When it was time for the fall pine nut harvest, the log or pole frames of wickiup houses like those at Bustos Wickiup Village awaited refurbishment. The Bustos site sat in a mature piñon forest but also in an area where a winter village might be nearby so that the stored nuts could easily be retrieved. Besides leaving houses there, people cached long hooked poles for pulling down cone-laden branches and big grinding stones and hullers for cracking open the nut meats. Even the circular rock storage facilities where people cached the harvested nuts had only to be refurbished and filled once again.

A cave or rocky ledge near a favored mountain hunting ground might shelter caches of arrow shafts made from the giant cane grass that grows in valley wetlands. A farmer in Willard, Utah, once found a storage pit on his land that contained more than six hundred small arrow points ready for use. They had been placed in a bag and buried in a small pit perhaps hundreds of years earlier. Ancient people also kept snare bundles in many places, along with baskets, bags, woven mats, stone axes, and digging sticks—just about anything that allowed them to go to work as soon as they arrived.

Tools for getting food were not the only items people cached. Archaeologists have found shamans' bundles, too, such as those from Humboldt Cave near Lovelock, Nevada. The bundles were little pouches holding pine pitch, ocher (iron oxide pigment), vegetal cakes that might have been medicines or prayer offerings, a stuffed weasel pelt with feathers in its mouth, and a host of other small objects. The so-called Patterson bundle, found in eastern Utah, is a shaman's curing kit with leather pouches containing individual doses of herbs, as well as a ball of pine pitch, pouches of stones, red ocher, a strand of deer dew claws, and much more.

As people used the landscape more and more fully, an inventory of metates and manos, the grinding stones used to mill plant foods, accumulated on the ground. Sometimes people stored them in the crotches of juniper and piñon trees or leaned them against tree trunks so that they could be easily spotted. In other instances, the coveted grinders were buried so that other people could not take them. Demonstrating the value of these tools, Southern Paiute consultants told the anthropologist Isabel Kelly in 1932 that they would make a new grinding stone only if an old one could not be found.

Not all early Great Basin people shared the same built environment. During the earliest times, when Paleoarchaic people first explored and perhaps colonized the land more than thirteen thousand years ago, few structures or caches existed, because people moved much longer distances in those days. But even then they cached valuable things such as the spectacular hoards of stone tool blanks uncovered in the Fenn cache near the intersection of Utah, Idaho, and Wyoming. This cache held eighteen pounds of superbly flaked blanks made of high-quality stone from quarries in all three states.

During the time archaeologists call the Archaic period, beginning about nine thousand years ago, people spread across the Great Basin and became more tethered to particular landscapes. The redundant use of places, relative to Paleoarchaic practices, stimulated greater use of a built environment. People constructed houses intended to be used again and again and invested in food storage facilities and animal traps. They made caches in caves and crevices and on ledges—places they could easily describe, remember, and locate.

By two thousand years ago, some parts of the Great Basin had literally become land filled with people. Distinctions between territories were strengthened. Where larger villages sprang up in some of the rich wetlands, some places and their resources became more exclusive. In a landscape with more neighbors, people exercised greater control over the built environment and began to hide caches of equipment and food from prying eyes.

## What to Eat and How to Get It

The cuisine of the ancient Great Basin was for the most part simple but probably less strange than the grislier stereotypes lead us to believe. On one hand, the Native diet strikes modern Americans as strong and bitter, yet on the other, its lack of fat and sugar makes it seem bland to modern sensibilities. Daily fare came mainly out of the stewpot. For most of antiquity, this pot was not ceramic, but a tightly woven, coiled basket whose contents were heated with hot stones from the campfire. One could boil a basketful of water this way in less than five minutes. Ingredients varied by season, but the stew often began with a base of flour made from seeds such as Indian rice grass, blazing star, saltbush, and native bluegrass, to name just a few. The cook might lace this mush with bits of meat, typically rabbit. Greens and seasonings such as thistle, peppergrass, and tansy mustard added spiciness. In the fall and winter, stews might be based on pine nut meal, one of the delicacies of the year. The basketry stewpot embraced the fruits of a landscape that

Figure 1.4. Great Basin people preparing roots and bulbs for cooking. Most ancient meals were stews cooked by placing food and water together with hot rocks—here heated in the fire in the center of the group—in watertight baskets.

Figure 1.5. A party in the northern Great Basin harvests bitterroot in the early summer. Roots were dug with digging sticks made of hardwoods such as mountain mahogany. Chipped stone tools known as crescents may have served to scrape the skin from the roots.

offered a variety of fresh foods rivaling that found in many modern supermarkets.

People also collected and processed starchy roots such as biscuit root, bitterroot, bulrush, cattail, and camas, which might be baked in the sand at the bottom of a campfire or in a rock-lined earth oven.

Figure 1.6. Situated at 7,600 feet in the Jarbidge Mountains of northern Nevada, this spectacular mountain sheep corral was used for thousands of years. Blinds were dug into the slope near the top center of the photograph, and a flattened butchering area can be seen inside the corral at the lower left.

Left in their skins or peeled with a stone tool with a concave sharp edge, roots could be wrapped in leaves and steamed. Many foods were best eaten raw, and people ate as they picked, nibbling throughout the day. Travelers might string dozens of bulrush and cattail roots the size of human fingers on a line that could be thrown over the shoulder or wrapped around the waist. This ancient gorp provided sustenance and served as tiny canteens, because starchy water made up three-fourths of the roots' weight.

Bread as we know it did not exist, but we have some evidence that people baked roots to a bread-like consistency. Curly dock seeds were sometimes pounded, soaked, and made into dough, then baked on coals. Cattail pollen was formed into cakes and cooked like tortillas on a stone slab.

Not all cooking was for immediate consumption. Desert fruitcake is a concoction made of whatever dried berries, meat, and seeds were available, mixed with animal fat to form long-lasting loaves. Roasted larvae of the pandora moth or the brine fly could also compose the base of desert fruitcake, preserving the superabundance of a highly nutritious food that was available only a few weeks a year. Desert fruitcake in all its variety provided a portable and concentrated form of energy and protein—an early version of the energy bars of today.

After stewing, roasting was the most common way of preparing meat. Bighorn sheep, mule deer, and of course pronghorn were common sources of roasted meats. Ancient people knew other prey animals as well as they knew the pronghorn for which they designed such clever traps. To attract bighorn sheep during the rut, hunters bashed two hollow logs together to simulate the sound of the rams' horns slamming together in mating contests.

Hunters also constructed traps. One such trap in the Jarbidge Mountains of northern Nevada was made of wood and stone fences built on a steep talus slope. Bighorn fleeing up rocky slopes easily outpaced their pursuers. But humans positioned above the trap could block the sheep's escape and force them to descend, where hunters popping up from blinds dug into the rocky slopes promptly shot them. These fences changed over time as hunters acquired new technology, shifting from the dart and atlatl to the bow and arrow. The adoption of guns did not make the trap obsolete; we find nineteenth-century shell casings from Henry rifles in the blinds. In one section of this mountain death trap, hunters even arranged the jagged stones to create a flat area for butchering their kills.

Despite thrilling images of big game hunts, small and medium-size mammals were the staples of the meat larder. Archaeological research shows that even before modern habitat encroachment, the supply of large animals was not endless. During some periods of antiquity, hunting kept their populations low enough that large game alone could not supply people's needs. The most commonly eaten desert meat in all of antiquity was rabbit, from both cottontails and jackrabbits, stewed or roasted on hot coals after the fur had been singed off. At a marmot roast in the summer of 1995 at Fish Lake, Utah, the Kanosh Band of Paiutes cooked the animals this way. The meat was dark and a bit greasy, but it was rich and filling. Ancient cooks gutted smaller animals such as squirrels, voles, and pikas by squeezing them and then made them into kebabs of a kind by inserting a stick into the body. Pieces of meat from larger animals were barbecued much as they are today. People made jerky to preserve meat.

Unlike most of us today, ancient people had to seek fat. Despite the variety of meats and the relatively high fat content of wild seeds and pine nuts, their diet was so low in fat that people actively sought this essential nutrient. Meat from wild game is almost completely lean except for fat under the skin and in the bones. Fat scraped from the skin bound together the ingredients of desert fruitcake. Fat skimmed from a boiling pot of bones might get a person through the worst days of winter. Left in, it certainly richened the stew.

Few people think that fish and deserts go together, but large wetlands exist in many Great Basin valleys, fed by mountain snowpacks and desert springs. They form mazes of contrasting habitats, from open ponds and spacious meadows to narrow channels lined by walls of rushes. The ponds and lakes offer a variety of sucker-type fish that people caught with nets or drove into schools that could be scooped out onto the banks during the spring spawn. Streams flowing from the mountains offered trout. Archaeology shows that people ate all kinds of fish, and in some places, such as Utah Lake and Pyramid Lake, Nevada, fish were a culinary cornerstone.

Perhaps the epitome of culinary opportunism and thoroughness appears in Lakeside Cave, on the edge of the Bonneville Salt Flats in Utah. For more than four thousand years, the ancient beaches outside the cave became occasional spectacles of superabundance. Whenever it rained, and during particularly wet centuries, water covered the salt flats. In the summer, when the winds were right, untold millions of drowned grasshoppers washed onto the beaches in ankle-deep windrows that could stretch for ten miles. People could collect tens of thousands of calories' worth of grasshoppers in a single hour, and each insect was 60 percent protein. This was a harvest no prudent forager would pass up. People carried the naturally dried and salted grasshoppers from the beaches and processed them in the cave. Coprolites, the dried human feces found in the cave, bristle with grasshopper parts. People must have known when conditions were right for these occasional jackpots and traveled to Lakeside Cave for the event.

Roasting and eating grasshoppers at Lakeside Cave in the early 1980s, my graduate student friends and I found the strip of white meat along their backs reminiscent of shellfish. We dubbed them "desert lobster."

The traditional diet was short on sweets, and one of the few sugars reported was aphid honey, deposited by the insects on plants such as cattails. People scraped it off with a flattened stick and ate

Figure 1.7. Fishers use nets anchored by stone weights to capture fish that will be dried and stored for future use. Fish taken during spawning season were a resource that could firmly tether people to a place. The tule boat on the water was used for harvesting a variety of marsh resources.

it. Although this diet may seem strange to contemporary Americans, people accustomed to it found the food rich and satisfying.

Maude Moon, a Goshiute Shoshone born in the late nineteenth century somewhere south of Wendover, Nevada, reported the change in her people's eating habits in the story "The Pickleweed Winter." Long ago, "Indians had everything they needed. They ate these things which grew on this earth…all kinds of seeds. This pickleweed, and also ones such as sunflower seeds, bunch grass seeds, rye grass, and just any kind, like *keppisappeh*, like wild onions, like Indian balsam, like carrots, like wild potatoes, like thistle.… During the winter, one ate all he wanted. It was over there at Big Springs, they called it the pickleweed winter. They ate it with pine nuts, they say. They ate it with jackrabbits. Times were good, they say.… But now you modern people, girls, and other modern Indians— they don't know anything. If they were gathered, they wouldn't eat them. They taste bad, they say. The sweetness has killed their mouths. They eat and drink canned sweet things. Only these taste good [to them today]. Indian food doesn't taste good anymore. It tastes too strong. It just tastes bad. It can't be swallowed. This is how it is."

**A Human Wilderness**

The ancients lived in the desert West with the nimbleness of long familiarity. They needed no street signs or maps because everything and every place had names and stories. Their languages held no word for "wilderness." The people marked no separation between humanity and nature, nor did they pose humanity against nature. The notion of "making a living" involved no distinction between work and play. There was harmony and balance, but these things were not static. The ancient people shaped their wilderness. They used it and sometimes even used it up. The balance they achieved was not a final state, but an unsteady relationship between human needs, beliefs, and the tyranny of circumstance.

The deserts and mountains of the Great Basin remain the last large wilderness in the lower forty-eight of the United States. Many of us can find a wilderness sense of place in the Great Basin, but in the ancient past, it was a human landscape, a human wilderness.

**Steven R. Simms** is a professor of anthropology at Utah State University.

Figure 2.1. Chaco Wash with Fajada Butte in the distance.

# Pueblo Farmers of the Chacoan World

*R. Gwinn Vivian*

On May 15, 1877, William Henry Jackson, pioneer photographer of the American West, discovered an ancient stairway behind Pueblo Bonito that allowed him to reach the cliff top for a spectacular view of the great house on his last day in Chaco Canyon. He had spent a week recording the major ruins in remarkable detail while camping in Chaco Wash beside "a few shallow pools of thick, pasty water." He ruefully commented that "the most important result…of this last discovery [the stairway] was the finding of a series of water pockets" containing "thousands of gallons of clear, cool, sweet water, a thing [he] had not seen for many days." The archaeologist Edgar Hewett once remarked that simply to get to Chaco Canyon, "the desert barrier must be crossed." Jackson's first attempt to reach Chaco from Fort Defiance had been thwarted because the Navajos "dread[ed] to visit it at this time of year on account of the well-known dearth of water."

But that desert barrier was penetrated many times in the past, beginning with nomadic Archaic gathering and hunting groups. They were followed by the first ancestral Pueblo Indians, who not only survived in Chaco Canyon but also thrived. By combining a knowledge of the land and the weather with human ingenuity, the early Chacoans adapted successfully to their new homeland, at least for a time.

This homeland—the Chacoan world—encompasses almost ten thousand square miles of the northwest corner of New Mexico, a physiographic region of the Colorado Plateau known as the San Juan Basin. Mountain ranges whose names reflect the Pueblo, Spanish, and Navajo heritages of past explorers and residents surround the lower-lying interior land. The inner basin is marked by one major topographic feature, Chacra Mesa, through which Chaco Canyon was cut in several stages beginning about three hundred thousand years ago.

The combination of the uplifted mesa and the channeling canyon created an oasis of sorts in an otherwise semiarid and almost featureless land. Rising as high as six hundred feet above the south side of the canyon, Chacra Mesa supports life zones not found on the valley floor. Slight differences in precipitation and temperature permit the growth of piñon, juniper, and other higher-elevation plants that provided humans with useful medicines and foods, including piñon nuts, along with habitat for upland-dwelling animals such as mule deer.

**Farming in Chaco Canyon**

The broad, flat canyon bottom was more suitable for farming. The runoff it receives from the east, north, and south made it one of the best-watered places in the inner basin. The major drainage, Chaco Wash, runs from east to west through the canyon, receiving additional discharge from two intersecting valleys on the south and from multiple short side canyons, or rincons, on the north (plates 14 and 15).

All this runoff comes from rain or melting snow. The amount of moisture entering the inner San Juan Basin and Chaco Canyon is conditioned to a large degree by the mountains surrounding the basin. These ranges form barriers to incoming storms and create a classic rain shadow effect. Continental atmospheric circulation systems, or "storm paths," also affect the quantity and timing

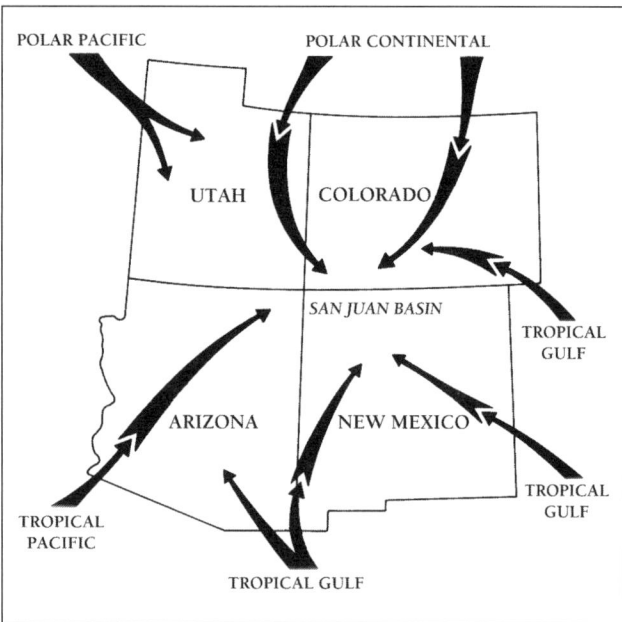

Figure 2.2. Storm paths that affect precipitation in the San Juan Basin.

of winter and summer precipitation in the basin. Southwestern Colorado lies near the southern limit of the snow-bearing winter path, and northwestern New Mexico sits on the northern edge of the summer rainfall path. This marginal positioning produces seasonal north-south differences within the basin and annual moisture differences if winter and summer storm paths shift slightly to the north or south. The ultimate effects of topographic and atmospheric factors on basin climate are highly variable rainfall and snowfall and an average annual precipitation of only 8.5 inches for Chaco Canyon.

The same factors influence temperature, which, with precipitation, is the most critical element for farmers. Seasonal temperature variations in the basin are extreme. Summers are hot, and winters can be cold to very cold. Record temperatures in Chaco Canyon have ranged from a high of 106 degrees F to a low of -38 degrees F. Corn usually is not damaged until temperatures drop below 30 degrees F, a "killing frost," but the plant does require between 110 and 130 days for full maturation. William Gillespie has noted that frost-free periods in valley bottoms such as Chaco Canyon are usually briefer than they are in uplands, shortening the growing season by thirty to thirty-five days. Fewer than half the years between 1960 and 1982 in Chaco Canyon had 100 frost-free days, and no year had as many as 150. As Gillespie observed, "it is hard to believe anyone would ever try to farm there."

I have spent much of my career searching for the reasons Puebloan farmers first came to Chaco Canyon and, more important, the reasons they stayed there for more than six hundred years. The first residents actually were not farmers, but Archaic gathering and hunting people who took seasonal advantage of the plants and animals on Chacra Mesa, a few seeps and springs in canyon alcoves, and the winter protection of rock shelters. The same amenities were attractive to the earliest farmers or part-time farmers who pushed into the canyon by at least 500 CE, lured by its runoff and adequate soils.

Recently, I had the good fortune to be involved in research that showed that the potential for farming in Chaco Canyon might have been enhanced for a while by an unusual geomorphic development that put cycles of arroyo cutting and filling in the canyon out of sync with regional channel dynamics. A natural sand dune formed at the western end of Chaco Canyon, near the great house called Peñasco Blanco, temporarily damming floodwaters flowing down the canyon and creating a broad, shallow lake. This dam effectively curtailed arroyo cutting in the canyon when such channeling started in the rest of the region in the mid-eighth century. Around 900 CE, the dune dam was breached, prompting a new cycle of arroyo cutting that lowered water tables and had other damaging consequences for Chacoan farmers.

Given the erraticism of Colorado Plateau climate, no ancestral Pueblo person could have depended upon just one farming method. Seasonal, annual, and spatial changes in moisture, cycles of erosion and filling, and fluctuating groundwater levels all forced Pueblo people to use multiple techniques and social means to reduce the risk of running out of food. Chacoans, like other ancestral Pueblo people, chose field locations that optimized yields and reduced the chances of crop failure. Within the canyon, good soils and exposure to the sun were important, but water remained the critical factor when they selected places to farm. Annual moisture was insufficient for dry farming under most conditions,

Figure 2.3. The remains of a prehistoric natural sand-dune dam across the mouth of Chaco Wash are still visible today.

so farmers planted crops where they would receive storm runoff or where they could be watered with captured runoff distributed through canals.

Runoff entering the canyon from the east, north, and south takes quite different patterns. Chaco Wash flows west through the canyon, carrying enormous quantities of water from many tributaries. When the wash is not entrenched, this water can spread thinly over much of the canyon bottom, including fields. During times of arroyo cutting, however, the water drains away quickly down Chaco Wash, providing no benefit for farming.

Two important sources of runoff, in very different ways, are the mesas bordering the north and south sides of the canyon. On the north, the mesa top above the cliff face is a wide bench with large expanses of bedrock and thin soil, whereas the mesa on the south presents a series of short, stepped and broken terraces, many with heavy soils. When moisture is sufficient to produce runoff, much of the flow on the north moves with considerable velocity across the bedrock and into short side-canyon arroyos that drain into Chaco Wash. Surface flow on the south runs more slowly and is absorbed into terrace benches, talus slopes, and the bottoms of side canyons. The differences are starkly reflected today on the two sides of Chaco Canyon. Almost every side canyon on the north has a narrow, deep arroyo, but there are virtually no arroyos on the south.

Because Pueblo farmers survived only by being acutely aware of their environment, they undoubtedly recognized the south side of Chaco Canyon as offering greater potential for simple farming techniques. There they found terrace benches well suited for small plots of corn, while the bottoms of side canyons were ideal for akchin (floodwater) farming. Sand dunes in places such as Werito's Rincon offered an additional micro-niche for planting beans, a method that was fully developed by the Hopi people to the west. And winter snow melts more slowly on the south side of the canyon, percolating deep into the soil to provide vital moisture for germinating seeds in the spring.

Tom Windes's long-term collection of rain-gauge data in and around Chaco Canyon indicates that there was an additional benefit to farming on the south of Chaco Wash. He discovered that when summer storms move from the southwest toward the canyon and meet Chacra Mesa, they are deflected into breaks or gaps in that barrier. This funneling, he found, tends to hold storms in the gaps for longer periods of time, resulting in much greater

Figure 2.4. A Navajo cornfield in Mockingbird Canyon, on the north side of Chaco Wash, in the early twentieth century.

precipitation. This could explain why Steve Lekson's "downtown Chaco," a dense concentration of great houses and small-house sites, is situated near the break called South Gap.

Some Chacoan farmers soon learned that if they were to take advantage of the full agricultural potential of the north side of the canyon, they had to capture water flowing off the mesa top and divert it to fields. I spent a full year in Chaco investigating ancient farming practices and located systems for capturing water in seventeen of the twenty-eight side canyons between Wijiji on the east and Peñasco Blanco on the west. My team and I also discovered water management features on the south side of the canyon, near Casa Rinconada and Peñasco Blanco, respectively. We were certain that undiscovered systems must lie in the rincons on the north side of the canyon.

The north-side farmers were remarkably consistent in the ways they collected, diverted, and spread the water that flowed off the slickrock and into short side canyons. They constructed earthen or masonry diversion dams near the mouths of these drainages to channel the runoff into canals. At the ends of the canals, they built headgates to further channel the floodwater onto fields. To ensure that all plants received equal water, the farmers gridded their fields into rectangular plots separated by low earth borders. Through the use of aerial photographs, we were able to identify several farms near Chetro Ketl that had individual garden plots averaging seventy-five by forty-five feet.

**Social Adaptations to Food Shortages**

Chacoans also used social strategies to deal with potential food shortages. Usually, ancestral Pueblo people responded to impending famine by moving to better farming areas. This happened occasionally at Chaco, particularly after 1080 CE, but people never fully abandoned Chaco Canyon at any time between 500 and 1150 CE. Instead, they seem to have successfully used ancient systems of organizing kin to good effect in certain canyon microenvironments. Evidence of two contrasting strategies may be reflected in the predominance of small-house sites on the south side of the canyon and of great houses on the north. Linda Cordell's analysis of Pueblo farming suggests a reason for this difference.

She proposed that Pueblo peoples practiced both labor-intensive and land- and time-intensive farming. In labor-intensive systems, well-coordinated and often large groups of workers constructed and maintained soil- and water-control facilities in fairly restricted areas. Land- and time-intensive systems were less technologically sophisticated, and smaller

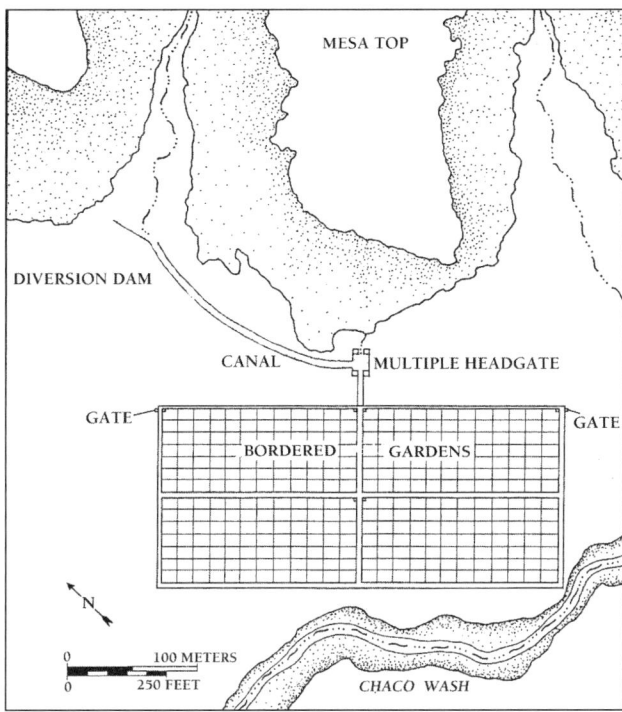

Figure 2.5. Plan of large gridded gardens using captured runoff from the north side of the canyon.

work groups often spent more time in larger areas of scattered plots practicing overplanting, multiple planting, fallowing, and shifting cultivation.

Chacoans living in small-house sites found the many micro-niche locations for fields on the south side of the canyon ideal for land- and time-intensive farming practices, and the dispersed nature of these sites suggests a social system based on extended families. On the other hand, the need for controlling and diverting large amounts of floodwater to gridded fields on the north side of the canyon would have demanded a more labor-intensive system. Efficiency may be gained by concentrating labor, and the few great houses along the northern margins of the canyon probably were organized in a way that permitted the clustering of larger social groups. Great-house farmers might have been organized like residents of the contemporary Tewa pueblos, where governing power is shared by two separate but equal groups.

The farming methods practiced by small-house groups were widespread and had deep Pueblo roots. Yet, the great-house populations successfully developed farming based on water control in Chaco Canyon over the course of three centuries. The same formality that characterized their north-side farms is evident in great-house planning, and I believe that changes in those plans over time may denote the responses of great-house groups to an altered environment.

The earliest ninth-century canyon great houses were larger than contemporaneous small-house sites, and their plans held embryonic hints of the "classic" D-shape so common in eleventh-century buildings. These structures were situated across from "funnel" zones in Chacra Mesa so that farmers could take advantage of increased rainfall and floodwaters entering the canyon through these breaks. By forestalling arroyo cutting, the dune dam below Peñasco Blanco also enhanced the value of bottomland farms. The earliest great-house farmers undoubtedly experimented with simple water-control devices, but these were probably designed more to spread water than to channel it. At the same time, small-house communities across the canyon practiced more traditional forms of planting.

By narrowing their farming methods, great-house farmers limited their options for adjusting to the breaching of the dune dam and a new cycle of arroyo cutting around 900 CE. Farmers in Chaco benefited from the regionwide gradual increase in moisture in the early tenth century, but entrenchment of Chaco Wash placed great-house floodwater fields in jeopardy because side tributaries eroded to the base of the main channel, lowering the water table and flushing water directly into the central wash. Great-house farmers did not experience the effects of this process immediately, but gradual headwater erosion of Chaco Wash ultimately threatened the farms near their clustered communities. Early expressions of the classic great-house plan, such as at Pueblo Bonito, seem to have become frozen in place for almost a century, a period Lekson describes as "the hiatus." Small-house farmers were less severely affected because hydrologic and geomorphic conditions on the south side of the canyon resisted arroyo cutting.

In the eleventh century, as precipitation increased and water tables rose, Pueblo people once again found Chaco Canyon an attractive place to farm. A return to channel filling might have been hastened and possibly even initiated when

Figure 2.6. Chaco Wash.

Chacoans built a masonry dam in the breached dune below Peñasco Blanco. Though it created a playa or possibly a shallow lake, it might not have been especially useful for crops. Instead, it may have been built to restore canyon bottomlands to a condition remembered from times past. This was a time when great-house groups invested heavily in water-control structures despite the occasional destruction of headgates and canals by flooding. The enlargement of older great houses and the establishment of several new ones, all reaching the architectural apogee of the classic great-house form, mirrored good times in Chaco. Small-house groups benefited from increased rainfall, too, continuing their practice of multiple micro-niche farming. Growing communities, however, may have begun to use up all suitable canyon farmland by the mid-eleventh century, and several new great houses, such as Pueblo Pintado and Kin Klizhin, were established outside the canyon's boundaries.

## The Final Decades

Movement beyond the canyon could also have been stimulated by a two-decade dry period that started around 1080 CE, though a stable water table and continued channel filling lessened its impact. These unsettling times were followed around 1100 by three decades of much greater precipitation. I believe that this may have lulled some Chacoan farmers into complacency, because it was accompanied by a surge in great-house construction. Apparently, though, not everyone believed that prosperity would continue, for between about 1080 and 1140, groups from Chaco Canyon established several large Chacoan great houses in the San Juan River valley, including Salmon Ruin and the West Ruin at Aztec. Chacoan concerns about the future were well founded, for a major drought that began around 1130 and lasted, with only a minor break, for fifty years underscored the harsh realities of depending on rainfall.

This event ultimately proved too taxing for the Chacoan farmers and triggered the abandonment of Chaco Canyon by the late twelfth century. By 1150, only small patches of stunted maize were being tended in the Chacra Mesa. This ribbon of life in the inner San Juan Basin had sheltered and sustained its Pueblo occupants for more than six hundred years, providing them a toehold of existence within a sea of desert uniformity. Though their isolation forced Chacoans to travel great distances—to the margins of the basin—for many of the resources they came to need, they had successfully farmed the canyon through tenacity, experimentation, and an occasional stroke of luck such as the dune dam.

The challenges faced by the people of Chaco sparked the genesis of their culture and then propelled them through centuries of often spectacular growth. In the end, the failing clouds, the drying winds, and the heat of summer prevailed, and the canyon returned to the way it was before.

**R. Gwinn Vivian** is curator emeritus at Arizona State Museum, University of Arizona in Tucson. A longtime researcher of Chaco Canyon and author of The Chacoan Prehistory of the San Juan Basin, he is particularly interested in how the people of Chaco adapted to their arid environment.

Figure 3.1. North rim of Mesa Verde, looking toward Sleeping Ute Mountain.

# Through the Looking Glass
## The Environment of the Ancient Mesa Verdeans

*Karen R. Adams*

For Alice in Wonderland, the task was easy. All she had to do was step through a magic looking glass into a wonderful new world that was not only clear but also in full color. Our archaeological looking glass, on the other hand, is fogged, and the images we peer at seem blurry at first. Plant remains and animal bones in archaeological sites usually are broken up, and plant fragments are often black from charring. Yet, as we archaeologists continue to assemble evidence from many sources, the haze in our looking glass clears, and we see in increasing detail the environment of the ancient Pueblo people of the Mesa Verde region. Little by little, we understand better how they went about growing crops, foraging for wild plants, and hunting animals.

We now know that for centuries farmers have considered the Mesa Verde country of southwestern Colorado and southeastern Utah to be a great place to settle down and raise a family. Well suited to agriculture, the region receives sediment carried by winds blowing regularly from the southwest, sediment that has formed thick layers of loose, nutrient-rich farmland. In addition, nature has satisfied many a Mesa Verdean farmer's need for enough moisture and frost-free days to enable crops to mature. For much of the time between 500 and 1300 CE (the Basketmaker III through Pueblo III periods), farmers raised corn (maize), beans, and squash, and they grew or traded for gourds. More recently, farmers in southwestern Colorado have experienced a century of successful dryland farming (farming by natural precipitation alone), particularly of splotchy red-and-white common beans similar to those grown by ancient farmers.

Despite the region's suitability for farming, we know that ancestral Pueblo people also experienced times of hardship caused by agricultural failures. More than once, they watched their crops wither from lack of rain or succumb to killing frosts. Sometimes hunger and malnutrition pitted groups against each other, resulting in warfare and emigration. The more we know about the environment of the past, the better we can understand population spikes and declines, the movements of people, and the ways they coped with food shortages.

The diverse biotic communities of the Mesa Verde region have long included many plants and animals that humans have found useful. Studies of preserved pack rat middens (nests) tell us that the present plant and animal communities of the American Southwest have been in place for the past four thousand years. The archaeological record verifies that ancestral Pueblo landscapes hosted many of the same wild plants and animals we see today.

Still, our ability to envision earlier landscapes has been inhibited by the appearance over the past century of invasive species originating on other continents. Tumbleweeds, summer cypress, and clovers, for example, now crowd out native plants. In addition, commercial farming, logging, and livestock grazing have cleared vast areas that once supported piñon and juniper woodlands and parklands of sagebrush and native grasses. Along with deep plowing and fire suppression, these historic activities have altered the proportions and, in some instances, the natural groupings of plants and animals. The role that ancient people played in altering their environment must also be acknowledged.

Figure 3.2. Ancestral Pueblo farmers built agricultural terraces such as these in Mesa Verde National Park to contain runoff and stabilize soil.

## The Modern Environment

Seven major biotic communities grace the Mesa Verde regional landscape. Each consists of a certain group of plants and animals that is affected by and adapted to local temperature, precipitation, and soil. From the valleys, at some four thousand feet in elevation, to mountain peaks above twelve thousand feet, these plant communities are characterized by, respectively, sagebrush and saltbush shrubs; grasses; piñon and juniper woodlands; Gamble oak scrubland; ponderosa pine and Douglas fir woodlands; spruce and true fir woodlands; and low-growing alpine tundra.

Piñon-juniper woodlands, expanses of sagebrush and saltbush, and grasslands abound in the southwestern corner of Colorado, where ancient human populations were once large. Although the region includes both major rivers (San Juan, Animas, La Plata, and Dolores) and smaller rivers and creeks (Mancos, McElmo, Piedra, and Yellow Jacket), the springs and ephemeral drainages have long been critical water sources for both animals and people. Over the centuries, Pueblo people also constructed check dams, water diversion systems, and reservoirs to make water more accessible.

All farmers know how critical moisture is to their crops. In the Mesa Verde region, mean annual precipitation ranges from 7.8 inches near Kayenta, Arizona, to more than 18.3 inches in and around Durango, Colorado. Moisture from snowmelt allows corn kernels to germinate and sustains tiny seedlings through the normally dry weeks of late spring and early summer. Later, sporadic summer rains spur rapid plant growth and ear development. On the Colorado Plateau, corn agriculture requires at least twelve to fourteen inches of annual precipitation to be successful, and some developmental stages of corn growth, such as pollination and grain development, especially need water.

Pueblo farmers learned long ago that they could

direct runoff from intense summer showers to their fields by aligning stones, dirt, and brush debris. They also placed their fields in locations best suited to receiving storm runoff, such as at the bases of gentle slopes. Sometimes they also hand-carried water from reservoirs ("pot irrigation"), especially to drought-intolerant crops such as squash.

Temperatures, too, play an important role in growing crops successfully. Most varieties of corn need at least 120 frost-free days and a minimum amount of summertime heat in order to mature. The latter is measured in "corn-growing degree day" (CGDD) units; at least 2,500 units are required during a growing season.

---

CGDD units are calculated by summing the difference between each day's average temperature and a set base temperature (50 degrees Fahrenheit). One CGDD unit is accumulated for each degree by which the average exceeds the base temperature. A minimum of 50 degrees and a maximum of 86 degrees have been set as thresholds below and above which corn crops will not thrive.

---

My colleagues and I have examined modern temperature and precipitation records for the Mesa Verde region to assess the locations best (and worst) suited for farming. Our research shows that despite being above seven thousand feet in elevation, Mesa Verde proper currently has enough CGDD units, frost-free days, and precipitation to farm successfully in most years, enhanced by the fact that the broad, sediment-covered mesas tilt gently south, toward the warming sun. The Yellow Jacket, Cortez, and Blanding areas, too, receive enough moisture and summer heat to grow corn, so it is no surprise that they were all densely populated in prehistoric times.

Tree-ring samples obtained from ancient roof beams, stands of very old living trees, and even older wood lying on rocky landscapes provide data on precipitation in the past. Dendrochronologist Matthew Salzer, working in the San Francisco Peaks area, near Flagstaff, Arizona, used bristlecone pine records to reconstruct periods of relatively higher and lower temperatures for the ancestral Pueblo centuries. He then combined the annual temperature and precipitation data to construct a two-thousand-year time line of conditions critical to agriculture. To add to this, Timothy A. Kohler and his associates from Washington State University recently assembled a broad range of environmental data, information about corn yields, and estimates of human population to form the basis of a sophisticated, multicentury model of periods of farming success and failure for the Mesa Verde region.

## Agricultural Success and Failure, Population Spikes and Declines

Farmers in the Mesa Verde region began growing corn, a Mexican import, sometime in the first few centuries of the first millennium CE (plate 11). Before long, it became both a dietary staple and an integral component in ceremonies. Why did this happen? Corn is unique among grain crops in having both large kernels and high yields. By planting a single kernel, a farmer could grow a plant that produced three hundred to six hundred kernels. Studies have shown that Pueblo farmers in the past century routinely set aside approximately 350 pounds of corn kernels for one person's annual consumption. Under favorable circumstances, three to four acres of land could yield enough corn to feed a family of three to four people for a year, assuming that wild plants and animals provided additional calories, critical vitamins and minerals, and protein missing from corn. Also, around 600 CE, Pueblo people began growing common beans. When added to corn in their diet, beans gave them all the amino acids of a complete protein.

After corn became a staple, the success and failure of corn harvests paralleled the rise and decline of the native population. By the 600s, population was on the increase, but a drought in the late 800s made dryland farming too risky. Throughout the 900s, Pueblo farmers seem to have hung on by tending small fields along drainages or by clearing fields at the bases of slopes to benefit from storm runoff. By the middle and late 1000s, while living in dispersed farmsteads and walking daily to their nearby cornfields, they were again reaping successful harvests. Between 1130 and 1180, however, drought again placed its curse on agriculture. We see its reflection in tree rings, in a reduction in

Figure 3.3. Pueblo people stored large quantities of corn, their most important food staple. The kernels safeguarded in this jar, excavated in Mesa Verde National Park, may have been specially selected seed corn.

house construction and remodeling, and in a rise in intergroup violence and strife. When that long dry spell ended, prosperity returned. By the late 1100s, population was growing and construction booming.

As the thirteenth century drew to a close, ending the Pueblo III period, the inhabitants of the Mesa Verde region once more experienced scant and unpredictable summer rains. This severe drought, which lasted from 1276 to 1299, spelled the end of eight centuries of Pueblo presence in the Mesa Verde and Four Corners region. Some communities suffered a violent end, and some people must have perished from the effects of malnutrition. Others emigrated to more promising places to the south and east to join relatives or friends and acquaintances.

## Gathering and Hunting

Over years of working in the Mesa Verde region, I have accumulated a list of important native plant foods that are found in the archaeological record or have been recorded in historic ethnographic literature. Ancestral Pueblo people collected tasty, calorie-rich piñon nuts, harvests of which can be abundant but often are sporadic and undependable. They occasionally gathered the reliable juniper berries, whose tartness (one species is used to flavor gin) requires some getting used to. Other nourishing wild foods harvested in large quantities included the seeds or fruits of weedy goosefoot, pigweed, purslane, tomatillo, tansy mustard, yucca, globe mallow, grasses, and cacti.

Ancient people knew when plant foods ripened and were ready to be collected. For example, starting in late spring or early summer, they gathered lemonade berries, rice grass grains, and tansy mustard seeds for immediate consumption and storage. Later in the summer, they likely gathered the tender and nutritious leafy greens of goosefoot, pigweed, and purslane. Although little evidence of the leaves and stems of these weedy plants has survived in the archaeological record, evidence for the use of their seeds is plentiful, and we assume that plants growing in farmers' fields and on midden piles produced copious seeds for harvesting.

As the growing season progressed, people ate a variety of grass grains, as well as serviceberries and sunflower seeds. By fall, everyone must have looked forward to gathering piñon nuts, if the harvest was

Figure 3.4. The yucca plant provided large, sweet pods for food and fibrous leaves for making baskets, sandals, and mats.

Figure 3.5. Rice grass produces abundant grains that ripen in late spring. Historically, Pueblo people cut the stems and held them over a fire, allowing the toasted grains to fall into a waiting container.

but during the warmer months, they preferred cooking outdoors. They parched wild seeds over a fire or ground them up to be cooked in pottery vessels as porridge or gruel.

Ancestral Pueblo people found many other uses for plants besides food. On occasion, they smoked wild tobacco, leaving burned tobacco seeds for archaeologists to find centuries later. Needing timbers for house construction, they chopped down juniper trees at lower elevations and Douglas firs and ponderosa pines on the higher mesas. They used piñon trees for building, too, but were aware that weaknesses in the wood reduced its value. Still, piñon and juniper, along with woody shrubs such as sagebrush, saltbush, bitterbrush, and mountain mahogany, provided fuel for cooking, heating, and light. And craftsmen no doubt tossed wood scraps left over from making household and agricultural tools into the fire.

In addition to knowing which wild plants to harvest, Mesa Verdean men and women knew when and where to find each one. We infer from the investment they made in architecture, good, and eating prickly pear fruit, one of the few sources of natural sugar. Because juniper berries remain attached to their tree branches for a long time, they could be gathered when needed. During the winter, women regularly prepared meals of stored corn and wild plants inside their dwellings, as well as from farmers' need to be near their crops during the growing season, that many lived in their dwellings and villages year-round. Because many of the wild plants they gathered thrive in disturbed habitats (we call such plants "weeds"), we assume that many harvests came from local agricultural fields and other places where daily living had disturbed the natural vegetation. Probably some members of each community occasionally traveled considerable distances from their pueblos to collect special plant resources.

Jonathan Driver, who has studied regional faunal records, believes that Mesa Verdeans hunted primarily local animals but occasionally sent small hunting parties farther afield. They obtained animal protein mainly from mule deer, jackrabbits, and cottontail rabbits, along with domesticated turkeys. Less often, hunters brought home bighorn sheep, pronghorn antelope, and elk. Although we find rodent bones in archaeological sites, they probably reflect a natural death, as well as hunting and trapping.

Figure 3.6. Mule deer, which thrive in pinon-juniper woodlands, were an important source of animal protein for Pueblo people.

As Driver has noted, the quantities of rabbit, deer, and turkey meat that Pueblo people consumed fluctuated over time. Basketmaker III and Pueblo I people, for example, relied on rabbits and large game and rarely ate turkeys, whereas later Pueblo II people regularly raised turkeys for their meat. After 1150, Pueblo people of the Great Sage Plain ate less big game and even more turkey meat. By then, they had overhunted deer near their settlements and diminished deer habitat through generations of tree cutting. Cottontails, however, continued to be hunted, for they apparently thrived in the brushy vegetation that grew on recovering fallow fields. The hunting pattern differed on Mesa Verde itself, where larger populations of big game continued to be available.

Like plants, animals provided useful products other than food. People wove blankets from turkey feathers and rabbit fur and made awls, needles, spatulas, ornaments, and flutes from turkey and deer bones. "Man's best friend," the dog, shared their accommodations, and favorite dogs were buried with due regard. Although today's fishermen flock to the region's rivers and man-made lakes, we have little evidence that ancestral Pueblo people caught or ate fish. This is puzzling. Part of the reason may be poor preservation of fish parts or the fact that fish caught and eaten on the spot leave no evidence back in villages or homes. But ancestral Pueblo people might simply not have eaten much fish.

### Anthropogenic Ecology: Humans' Effects on Their Landscape

For eight centuries, inhabitants of the Mesa Verde region farmed, foraged, hunted, chopped down trees, and did much else that affected their landscape. From time to time, intentionally or not, they also set fire to portions of forests. We refer to such human influence on plant and animal communities as "anthropogenic (human-caused) ecology."

Archaeologists Tim Kohler and Meredith Mathews were among the first to report evidence of this process, from Pueblo I village sites around Dolores, Colorado. In the ancient plant remains, they observed a shift over time in fuelwood use, from piñon and juniper trees to shrubby plants and cottonwoods. That is, as preferred trees were used up, people sought alternative fuels still locally available.

With Vandy Bowyer, I took a similar look at the Pueblo III plant record of the Sand Canyon locality near Cortez, Colorado, and concluded that although

Figure 3.7. Bean farming thrives today, as it did centuries ago, in the Montezuma Valley, north of Cortez, Colorado.

Pueblo people had cut many trees and opened up land for agriculture, portions of piñon-juniper forests still remained within walking distance of their pueblos. That people consumed less corn in the late 1200s might reflect more than the onset of drought: the fertility of the land might have diminished, too, reducing its ability to produce enough food. It is clear that by this time, preferred foods, including wild game, had become scarce and people turned to foraging for wild plants. At Salmon Ruin, to the south, we know that hungry residents resorted to eating leftover corncobs, normally used only as tinder or fuel, as well as juniper bark, broad yucca leaves, animal bones, and insects.

## Closing Thoughts

Reconstructing past environments can be a challenge. Many facets of life affected by the environment simply are invisible to the researcher's eye—a maverick late spring frost or an early fall frost, for example. Did eight centuries of intermittent farming diminish nutrients in corn fields? Can we even document such trends after modern farmers have cultivated the same land? To complicate matters, it is difficult to assess the effects of historically introduced weeds that have invaded modern landscapes.

Despite such difficulties, we archaeologists continue to fine-tune our knowledge of the Mesa Verde region's past environment. Our multiple lines of evidence include plant and animal data preserved in archaeological sites; reconstructions of ancient weather; knowledge of modern plants and animals and of the best places to farm; and the responses of living indigenous groups to their environmental problems. As our archaeological looking glass continues to clear, the ancient Mesa Verde landscape comes into sharper focus—not quite like Alice's experience, perhaps, but satisfying nevertheless.

**Karen R. Adams** is an archaeobotanical consultant who has analyzed and interpreted plant remains in the American Southwest since the early 1970s. Her research interests include the contribution of plants to ancient subsistence and other daily needs, and shifts in plant usage over time. She is currently a research associate at Crow Canyon Archaeological Center and lives in Tucson, Arizona.

Figure 4.1. Cave 7, in Whisker's Draw, southeastern Utah, being excavated in 1893 by members of the Hyde Exploring Expedition.

# Ancient Violence in the Mesa Verde Region

*Kristin A. Kuckelman*

Near a small, uninhabited cliff overhang in southeastern Utah, now called "Cave 7," sometime during the late Basketmaker II period (200 BCE–500 CE), almost a hundred men, women, and children perished in a single assault—the earliest known episode of warfare in the northern Southwest. Someone buried their bodies, along with some of their belongings, under the overhang. In 1893, Richard Wetherill, rancher, archaeological explorer, and leader of the Hyde Exploring Expedition, discovered the remains of the victims but found no evidence of where they had lived or why so many people had congregated at this place. Basketmakers normally lived and traveled in small family groups; perhaps this large group was wintering together or had gathered for an important religious or social event. Why these people were killed and who the killers were will forever remain mysteries.

About 1280 CE, near the end of the Pueblo III period (1150–1300), attackers killed many residents of Castle Rock, Sand Canyon, and Goodman Point pueblos and ended the occupation of these impressive, stone masonry villages in what is now southwestern Colorado. Within a few years of these attacks, Pueblo people no longer lived in the Mesa Verde region.

Archaeologists from Crow Canyon Archaeological Center found evidence of the assaults during excavations over the past thirty years. By carefully excavating sites of attacks and massacres such as these, other archaeologists and I have learned much about violence among the ancestral Pueblo people of the Mesa Verde region.

Violence is a part of human society. Although the Pueblo residents of this region interacted peaceably much of the time, violence did erupt in forms ranging from interpersonal conflict, such as battery and murder, to large-scale warfare. Many violent acts leave a trail of evidence, and important clues to the nature, scale, and timing of warfare show up in the archaeological record. Structures that have defensive architectural features or that were constructed in defensible locations reveal that the builders feared being attacked. Intentionally burned buildings, large numbers of weapons, rock art images of warriors or battle scenes, and oral accounts of warfare related by the descendants of ancient Pueblo people suggest that conflict took place. The skeletal remains of the victims themselves provide the most direct and telling evidence of violence: skulls and other bones that were fractured around the time of death ("perimortem"), bones with embedded projectile points, skeletons in awkward positions where bodies were dumped into structures, and bones that were scattered because bodies were left unburied on rooftops or on the prehistoric ground surface. These clues give archaeologists a valuable record of violent events.

**Defensive Structures and Defensible Locations**
Residents of the Mesa Verde region took a variety of precautions against attack. During some time periods, family groups built stockades or palisades around their farmsteads. We have found several such sites dating from the 600s and several others built in the 1000s. But because archaeologists tend to focus on refuse areas and main buildings—whereas people built stockades around the perimeters of their farmsteads—many stockades may remain undiscovered.

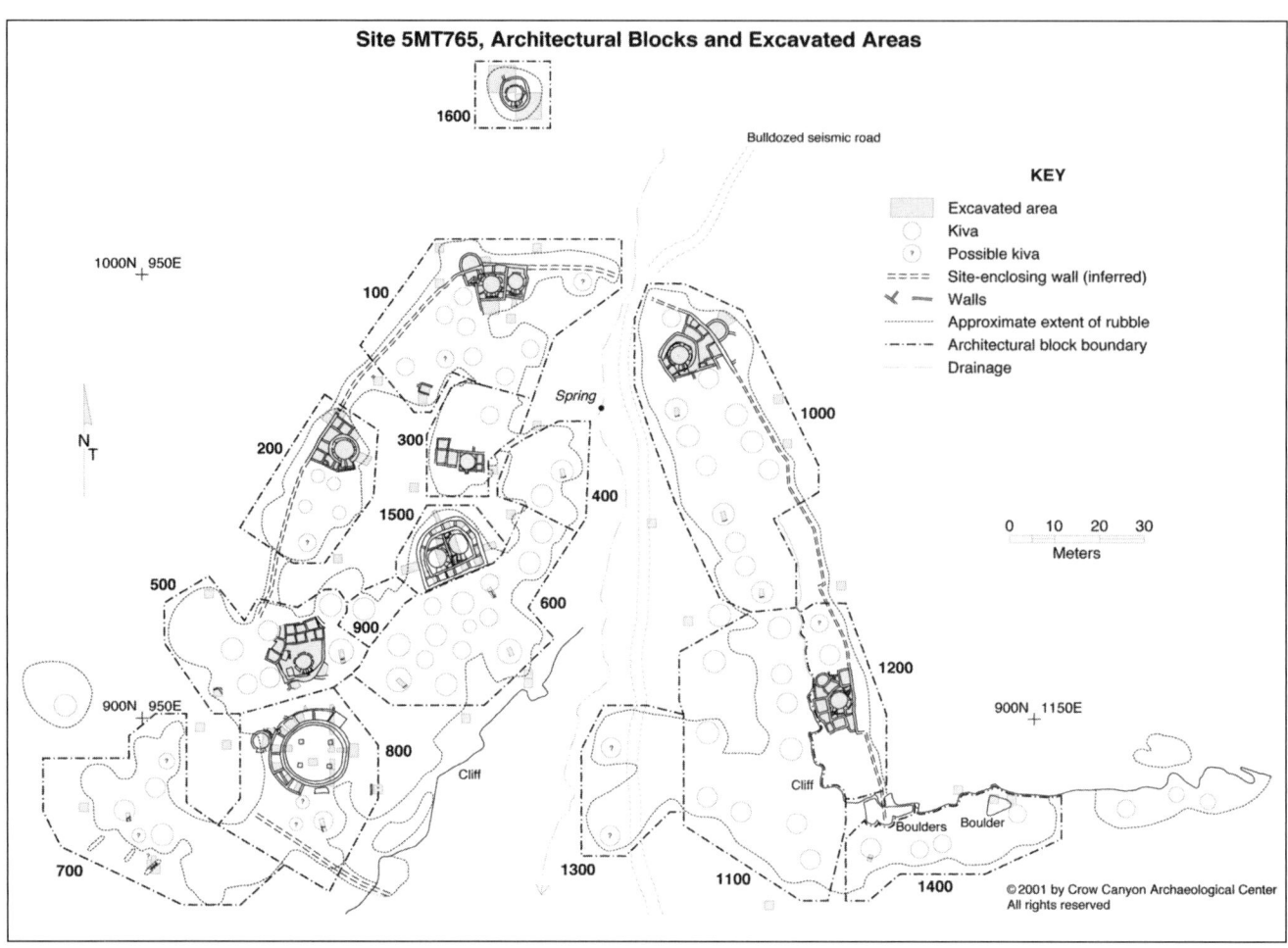

Figure 4.2. Plan of Sand Canyon Pueblo, with its defensive enclosing walls.

The aboveground portions of stockades deteriorated long ago, and we are unsure of the exact procedures used to construct these formidable walls. We do know from excavated post holes and post remnants that people invested a great deal of labor in building stockades. We think that the posts were interlaced horizontally with willows or pliable tree branches and the entire construction coated with mud or adobe. In the 1200s, people built stone walls to enclose villages such as Sand Canyon (plate 5) and Goodman Point pueblos. These massive, durable constructions defended the villages even more effectively than timber stockades. Some stockades and stone enclosing walls have additional defensive features such as strategically placed peepholes and few access openings.

During the Pueblo III period (1150–1300), people of the Mesa Verde region built stone masonry towers of various shapes and sizes that would have been useful in the event of an attack. Most towers were one or two stories high and were built with beautifully shaped and "pecked," or dimpled, blocks of sandstone. People could have escaped from some kivas (residential structures typically built underground and entered through roof hatchways) through tunnels connected to towers, which protected them from becoming trapped during a surprise attack. Many towers were freestanding; others adjoined adjacent buildings or the outside of a village-enclosing wall, much as military forts worldwide have been configured for centuries. From these buildings, lookouts commanded a panoramic view of the surrounding landscape and were better able to spot and defend against approaching enemies.

In the mid-thirteenth century, in the final decades before the region was completely depopulated, many people relocated from their small, scattered

Figure 4.3. The placement of its doorway at the edge of a cliff made access to Hovenweep Castle, a thirteenth-century building in southeastern Utah, extremely difficult.

farmsteads to nearby canyon rims, cliff overhangs, and other defensible places. There they constructed villages around or near their water sources. This clustering of people could have been both an offensive and a defensive, safety-in-numbers strategy. To restrict access and enhance safety, villagers constructed many single- and multiple-story buildings without doorways in the lowermost story—those ground-level rooms could be entered only through a hatchway in the roof. They erected some buildings at the very edges of canyon rims and placed the only doorway in the wall facing the canyon, so that it could be entered only from the rooftop of a lower structure inside the canyon.

It was during this time that the inhabitants of the Mesa Verde escarpment built the now internationally renowned cliff dwellings of Mesa Verde National Park. These settlements, sheltered beneath cliff overhangs, boasted excellent defensive properties: they were difficult to see from a distance, complicated to reach, and strategically dangerous to attack (plates 14 and 21). In addition, most of the overhangs contained natural springs, which greatly enhanced the residents' ability to withstand prolonged attacks or sieges. The residents of cliff dwellings protected themselves even further by storing food in granaries in small cliff alcoves and on ledges that were difficult to reach. They used some small overhangs strictly as defensive refuges by constructing a masonry wall across the front of the alcove and entering the haven by means of a ladder that could be pulled up to prevent attackers from following.

During some assaults, attackers burned buildings in an attempt either to kill the occupants or to destroy their homes. But Pueblo homes built of

Figure 4.4. Eagle Nest, a defensively situated cliff dwelling in Ute Mountain Tribal Park.

Figure 4.5. This remote granary in Grand Gulch, Utah, was well protected by its location on a ledge high off the canyon floor.

stone masonry did not burn readily; only the wooden timbers, which formed part of the roof, were flammable. Although some archaeologists believe that many buildings in the Mesa Verde region were burned during enemy attacks, the evidence indicates to me that most burned structures were set afire not by attackers but by the residents themselves, probably as part of a ritual "closing" of the building.

## Weapons, Rock Art Images, and Oral Accounts

In warfare, the ancient Pueblo people used spears, wooden clubs, stone axes hafted on wooden handles, and bows and arrows. But they used these objects for hunting and other purposes as well; they had no weapons exclusively for warfare. They did, however, make shields of basketry or hide for combat only, so the presence of these items or depictions of them in rock art point to conflict in the region. Warriors struck fatal or debilitating blows to the heads of their enemies with shock weapons such as axes and clubs. They used bows and arrows for striking from a greater distance, although fewer of the wounds caused by arrows would have been fatal.

Shortly before the Mesa Verde region was depopulated late in the thirteenth century, Pueblo people began to depict violent scenes and warriors

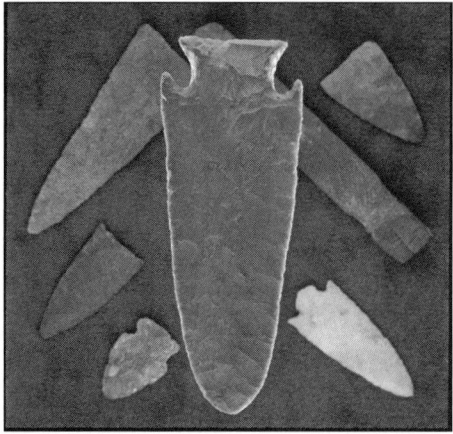
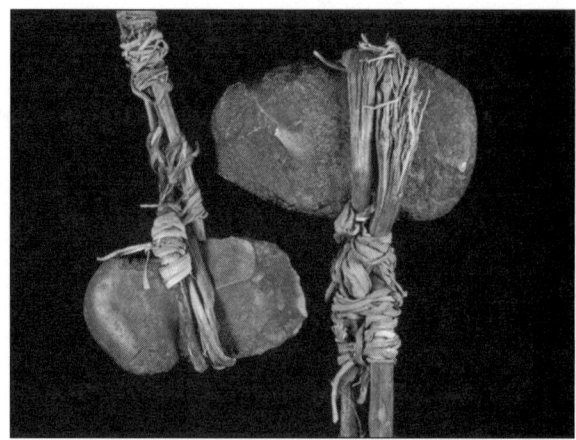

Figure 4.6. *Left*: a five-inch chert blade found lodged between the ribs of a Cave 7 skeleton; *right*: a pair of axes, also excavated in Cave 7.

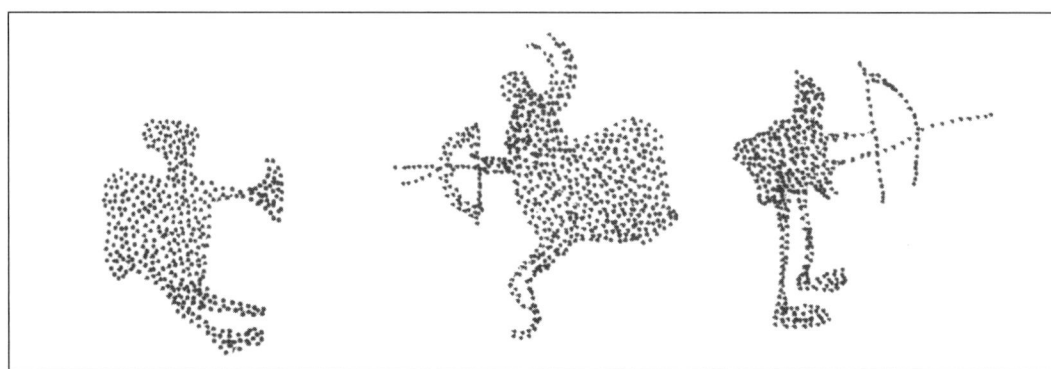

Figure 4.7. Drawings of petroglyphs on Castle Rock depicting a combative scene.

with shields on the faces of cliffs and boulders. An image of armed conflict at Castle Rock Pueblo was perhaps made to symbolize the violence of the times or to record the massacre that occurred there.

Oral accounts of specific prehistoric events are seldom available to archaeologists. In the mid-1800s, however, a Hopi elder related a centuries-old account of an attack on an ancestral village identified as Castle Rock Pueblo. Some details in his account match the archaeological remains found during Crow Canyon's excavations at the site. He said, for example, that many of the village residents were killed, causing the occupation of the village to end, and that the attack took place during mass migrations from the region. Because the elder relating the story was Hopi and because he said that the surviving villagers moved to the Hopi mesas in what is now northeastern Arizona, it is possible that some modern Hopi people descend from residents of Castle Rock Pueblo.

## Wounds and Other Skeletal Damage

Many victims of interpersonal violence and large-scale attacks were struck on the head forcefully enough to fracture the skull. Some died as a result, but the presence of healed fractures on numerous skulls proves that others survived the injury. The uniform size and shape of many of these fractures indicate that they were inflicted by a single type of weapon—probably a small, sharp-bitted stone ax hafted onto a wooden handle.

Typically, ancestral Pueblo people buried their dead carefully, placing the body in an abandoned building or burying the remains either in a midden

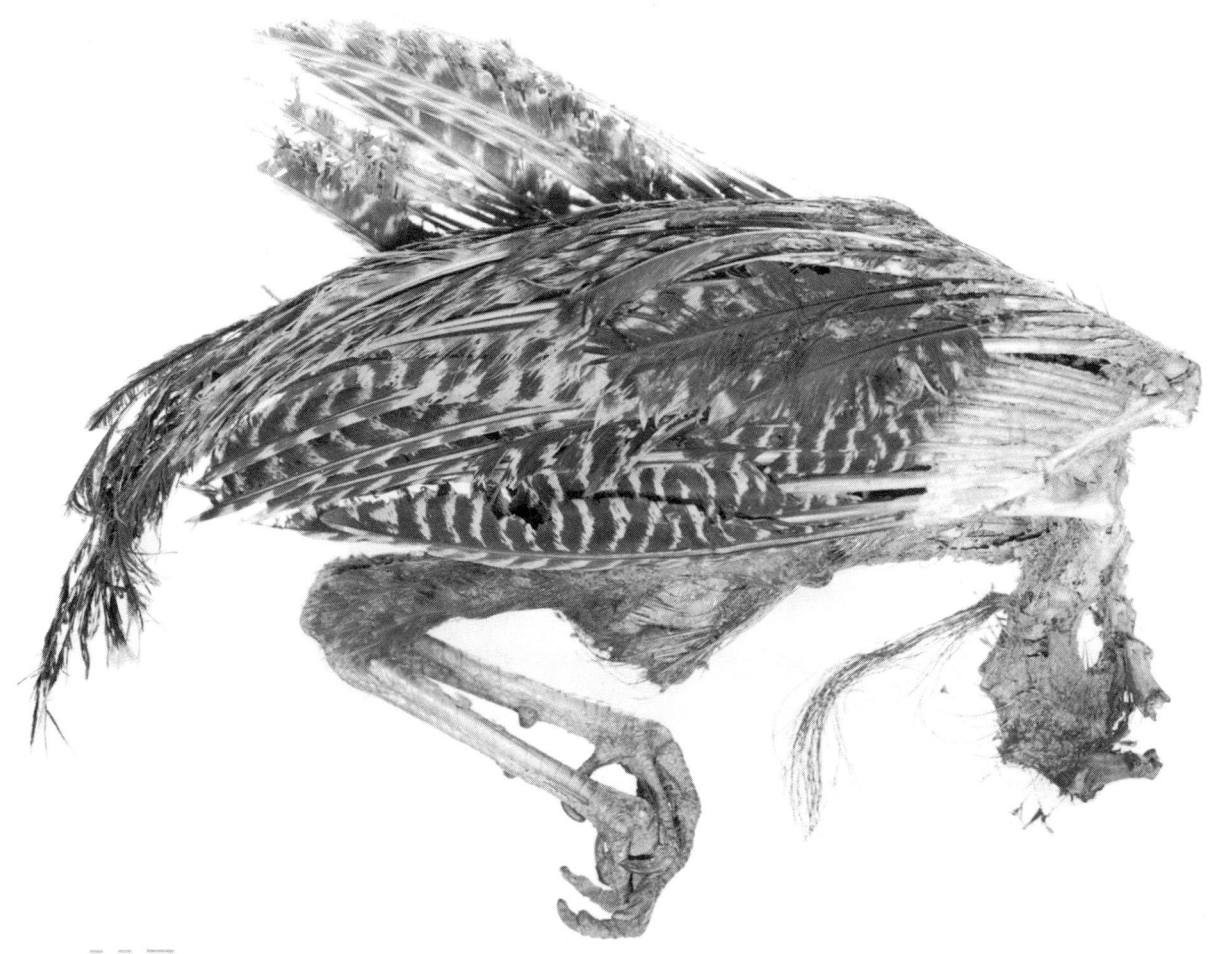

Figure 4.8. This mummified turkey was found in a rockshelter site just south of the Mesa Verde region. By the seventh century CE, Pueblo people were regularly keeping turkeys, and by the twelfth century CE, they were an important food source.

(refuse) area or beneath the floor of a room that was still in use. The bodies of many men, women, and children who died at enemy hands, however, were not buried—they were left wherever they fell or were dropped through a nearby roof hatch or doorway, either by the attackers or by relatives or friends of the victims.

Warriors sometimes scalped or dismembered their victims. After the fatal attack on Castle Rock Pueblo, victors probably engaged in anthropophagy (cannibalism). The practice of anthropophagy in the Mesa Verde region might have been a means to survive famines associated with severe droughts or other environmental conditions that destroyed the maize crops on which ancestral Pueblo people depended heavily for survival.

**Causes of Violence**

What situations and conditions caused the ancestral Pueblo people of this region to act violently? Although determining causes of human behavior is vitally important to understanding ancient cultures, we are hard-pressed to discover these causes from archaeological remains. It is probably safe to assume that personal assaults resulted from a great variety of disagreements between individuals or families and that people engaged in warfare only as a result of more serious or widespread problems.

Competition over food and other resources has precipitated warfare in many cultures worldwide. In the Mesa Verde region, numerous sites with evidence of anthropophagy were inhabited during the mid-1100s or the late 1200s, both of which were times of serious drought. The earlier of these droughts lasted from 1140 until 1180, and the latter, often called the "Great Drought," parched the landscape from 1276 until 1299.

During the early and mid-1200s, residents of the Mesa Verde region relied heavily on maize and on meat from domestic turkeys. Recent evidence from Sand Canyon Pueblo indicates that when the climate deteriorated in the late 1200s, maize crops failed and turkey flocks, which were fed maize, were decimated. People were forced to search out and eat both a wider variety and less preferred types of wild plants and animals and to compete with residents of other communities for these dwindling resources. Social, political, or religious problems might have caused conflicts during this time, but increasing violence in other parts of the Southwest and of the entire continent during the late 1200s and early 1300s suggests that the cause was a wide-ranging one such as a climate shift, instead of societal problems particular to the Pueblo people of the Mesa Verde region.

**Who Were the Attackers?**
The archaeological evidence of violence and warfare that we have so far unearthed in the Mesa Verde region consists mostly of the remains of the Pueblo victims and their villages. Although we have been unable to determine the identity of the aggressors of any particular attack, only a few non-Pueblo artifacts have been found at the scenes of these attacks, which might indicate that the perpetrators were other than Pueblo. These artifacts consist of a few stone arrow points of a style suggesting a southern Utah origin. I believe that the invaders were warriors from other Pueblo communities in the vicinity. In addition to the negative evidence just mentioned, non-Pueblo groups near the Mesa Verde region appear to have been too small to have successfully attacked pueblos as large, well fortified, and well defended as Castle Rock, Sand Canyon, and Goodman Point pueblos, let alone to have inflicted such heavy casualties. At the time of the final attacks on Sand Canyon and Goodman Point pueblos, they were the two largest settlements in the region. Severe shortages of food or water might have provided ample motivation for Pueblo people in the Mesa Verde region to go to war against neighboring Pueblo communities.

**Effects of Violence**
Violence and warfare affected the lives of the ancestral Pueblo people of this region in many ways. Warfare sometimes determined where people chose to locate their villages and influenced their social, political, economic, and religious systems. People built defensible villages near their water sources. The clustering, or aggregating, of people must have brought new problems, such as a higher incidence of infection and contagious diseases. The increased traveling required for people to tend crops, hunt, and collect firewood heightened their exposure to ambush. When men left home on communal hunts, they left their families more vulnerable to attack. The population shrank as men, women, and children fell victim to violence. The loss of children and of women in their reproductive years reduced the population's ability to flourish; the deaths of warriors eliminated providers critical to survival. In big attacks like those on Castle Rock, Sand Canyon, and Goodman Point pueblos, the unique or specialized knowledge of particular groups was lost. Most important, intensifying violence was undoubtedly one reason people decided to emigrate from the region.

**Conclusion**
Among the ancestral Pueblo people of the Mesa Verde region, the incidence of interpersonal violence, the threat of warfare, and the eruption of large-scale attacks waxed and waned over centuries of changing societal and environmental conditions. Because archaeologists have excavated only a tiny fraction of the archaeological remnants left by these people, we cannot determine how widespread the violence, raids, and large-scale assaults were at different times during the Pueblo occupation of the region. Casualties were heaviest when droughts or other environmental hazards diminished or destroyed maize crops and heightened competition for food and water. No doubt, social, political, and religious issues also caused conflicts, but these types of causes leave few traces for archaeologists to find.

Figure 4.9. Castle Rock.

Conditions such as drought, resource depletion, and overpopulation have recurred throughout human history and continue to plague people today. Although drought is unpreventable, resource depletion and overpopulation might be halted through long-term, thoughtful planning. Also preventable is overdependence on one resource, such as maize, which might have been a fatal flaw in the subsistence strategy of the ancestral Pueblo people of the Mesa Verde region. Even though conflict is part of human society, we might be able to reduce its frequency by learning from past incidents of violence and warfare.

**Kristin A. Kuckelman**, senior archaeologist and research publications manager for the Crow Canyon Archaeological Center, has spent more than three decades conducting field research in the Mesa Verde region. She has written many reports, journal articles, and chapters for edited volumes on the ancestral Pueblo people.

Figure 5.1. Kenneth Chapman and Eleanor Johnson tracing cavate pictographs in 1915.

# Carved in the Cliffs
## The Cavate Pueblos of Frijoles Canyon

*Angelyn Bass*

Peering into one of the hundreds of cave dwellings carved deep into the cliffs on the eastern flank of the Jemez Mountains, visitors try to imagine the lives of the Pueblo people who built and lived in them centuries ago. We can all appreciate the hours of labor people put into creating them, speculate on the tools they used to carve them, marvel at the images they painted on the walls and ceilings, and imagine generation upon generation of householders weaving cloth, making pottery, and grinding corn.

Though archaeologists have studied the "cavates" for more than a century, we still do not know exactly why and how they were built or used. Cavate pueblos display most of the features of the large, free-standing masonry pueblos and were clearly used as residences—but why did the Pueblo people build dwellings in the cliffs instead of stone-walled pueblos in the open? The presence of cavates becomes even more curious when, as in Frijoles Canyon, they are situated next to masonry pueblos where people lived at the same time.

More than a thousand cavates dot the Frijoles Canyon cliffs, and countless more nestle in the canyons of the Pajarito Plateau (plate 16). Cavates and their interior features give us a rare glimpse into the daily lives of Pueblo people of earlier times. Cavates were dwellings of the ancestors of modern Pueblo people who still make their homes in the Rio Grande Valley. The Tewa word for "cave dwelling" is *t'ová tewha*, which also translates roughly as "old or crumbling village against the wall."

Cavates are unique in the Pueblo architecture of the American Southwest. Not only are they excavated into the rock (hence the name *cavate*), but also their interiors hold beautifully preserved features that people used for food preparation, storage, and weaving. Some cavates have painted and incised plaster and petroglyphs that hint at formal ceremonial use.

Today the cavate pueblos appear as a multitude of partial and complete chambers. Originally, some reached four stories high and encompassed both cavates and exterior rooms built partly or entirely of stone masonry. Little is left now of the exterior masonry, which has mostly collapsed. At Long House in Frijoles Canyon, for example, all that remains of the exterior portions of this once large cavate pueblo are stone foundations at the base of the cliff, plastered areas on the cliff face that were the back walls of rooms, and horizontal rows of empty sockets that once held roof beams. To help visitors imagine what the now collapsed walls once looked like, Talus House, just down canyon from Long House, includes a reconstruction built in the early part of the twentieth century by Kenneth Chapman, an anthropologist, and a Tewa Indian crew from San Ildefonso Pueblo. They built Talus House on original masonry foundations uncovered by Edgar Lee Hewett, the first archaeologist to conduct extensive excavations in the canyon.

From these past excavations, as well as more recent surveys and studies, we know that Pueblo people began constructing cavate pueblos in the mid-1200s CE. They continued building them until the mid-1500s, with a peak in use in the 1400s. We calculated this chronology from different sources: analysis of pottery found inside the chambers and

Figure 5.2. Cavates along the north cliff face of Frijoles Canyon.

on nearby talus slopes, tree-ring dating of wood excavated from cavates, and archaeomagnetic dating of a few hearths. In Frijoles Canyon, the dates reveal that the cavates were occupied at the same time as the large, freestanding masonry pueblo of Tyuonyi, though we are still unsure of the exact relationship between the two. The chronology of cavates in other parts of the Pajarito Plateau is less well known than that of Frijoles Canyon, but most were probably built during the drought-punctuated 1400s and early 1500s. As in Frijoles Canyon, the other cavate villages lie close to large, Classic-period masonry villages and sources of water.

**Cavate Construction**

Cavates of the Pajarito Plateau were carved into layers of volcanic ash known as Bandelier Tuff. These layers formed during two separate eruptions and ash flows from the Valles Caldera volcano, the first approximately 1.6 million years ago and the second about 1.2 million. Portions of the tuff are weakly cemented, especially at the junction of the two ash flows. Pueblo people took advantage of this weakness by excavating cavates into the glassy ash where the flows meet. To get afternoon sunlight during the cold plateau winters, they cut caves into south- or southeast-facing cliffs. North-facing slopes and canyon bottoms, in contrast, are shaded in winter by mid-afternoon. From the deep striations and gouges in most of the cavate ceilings, we can tell that their builders pecked, carved, and chiseled out the soft tuff with tools such as digging sticks and sharpened stones, cutting progressively deeper into the cliff-face bedrock until the room reached a desired shape and size.

Many cavates are single chambers, but some are connected by doorways. The Pueblo people occasionally built masonry walls and partitions inside the cave rooms, as well as front walls and entrances to rooms cut in the rock. The roughly hewn tuff blocks are bedded against each other, with mud and small rubble pressed into the voids between the stones. Though the coursing is random, the stonework forms nearly vertical columns at corners and jambs. Mud mortar was used for pointing, for bedding and head joints, and as a plaster on some surfaces. Unfortunately, because of their rough construction and exposure to the elements, only a few of the masonry walls remain in place today. Most have collapsed, leaving behind just traces of the building materials as clues.

Cavate shapes and sizes vary considerably. Most are semicircular in plan, but some are rectangular. The builders usually made the ceilings hemispherical rather than flat, almost certainly because they knew that this shape provided structural stability in

Figure 5.3. An intact partition wall of a cavate.

the soft tuff. In the same way an arch or an egg has tremendous strength because of its curved shape, cavates with hemispherical ceilings can withstand the weight of the rock above, making them less likely to cave in. Though some cavates have collapsed, this appears to have been due to their location at naturally occurring and intersecting fractures, rock falls, and centuries of weathering and erosion.

## Architectural Finishes and Embellishments

After carving a cavate, many builders intentionally sooted its interior. The soot not only created a uniform interior color and texture but also hardened and coated the grainy tuff surface to help prevent it from crumbling. After blackening the ceiling and walls, people scraped the soot off the lower half of the walls (to provide a solid bonding surface) and then plastered that portion and the floor. Cavate plaster was specifically crafted of clay, silt, and sand and often included ash and organic matter to help it cohere.

Plaster served several purposes. As a protective coating, it shielded the tuff from moisture infiltration and erosion. It also insulated the room, controlled dust, and provided a smooth surface for painted or incised designs. Cavate dwellers finished their floors, too, with earth plaster, creating a hard, even surface on which to sit, sleep, and work at mealing bins, hearths, and looms. Often the wall plaster is an extension of the floor plaster, indicating that builders created floor and wall surfaces at the same time.

Most cavates have an interior dado, or band of plaster covering the lower half of the walls. This plaster band generally measures approximately three feet high—about eye level if one is sitting on the floor. Above the dado, the wall is sooted tuff. On rare occasions, cavate residents fully plastered a room's interior, including the ceiling. Some cavates have many layers of plaster (one has more than thirty), revealing that people frequently refinished the walls, especially at the doorway and dado levels, where there was frequent wear and tear from access and day-to-day activities taking place on the floor. When they replastered the dado, they often started the new layer slightly lower on the wall, leaving a dark band of the older plaster exposed at the top. Frequently, this dark band is decorated with incised images. A few rooms have stylized anthropomorphic and zoomorphic figures or geometric designs painted on the walls, and some have petroglyphs carved into the sooted walls and ceilings.

Since at least the 1200s, Pueblo people have plastered the walls of rooms and painted portions of kivas with murals. At Bandelier, the plastered walls of some cavates display wall paintings. Snake Kiva, excavated by Hewett in 1909, is so called because a large horned serpent, or Awanyu, is painted along the curving back wall. In the 1930s, fragments of this painted surface plaster were flaked away to reveal another mural beneath it: an anthropomorphic head, in profile, with a feather headdress, painted in red, yellow, and white. Several other exposed patches of yellow and red paint along the wall suggest that this earlier mural extends all the way around the cavate. Such layered murals are suggestive of other complex narrative murals found at Pueblo sites such as Awatovi, Pottery Mound, and

Kuaua. The murals at these villages were painted for specific ceremonies and plastered over at the completion of each one. Perhaps the Snake Kiva mural was covered for the same reason.

## Cavate Use

We cannot say exactly how cavates were used because their functions might have changed over time, but we have many clues. In the 1980s, the archaeologist H. Wolcott Toll identified three broad categories of cavate rooms as habitation, storage, and special purpose. He based this classification on two important characteristics: a cavate's size and the features found inside and around it. By now, archaeologists have examined more than a thousand cavates in Frijoles Canyon and estimate more than three quarters of them to be for habitation or dwelling, with sizes averaging between twenty-five and ninety square feet. Fewer than a quarter are smaller storage rooms, measuring between three and twenty-two square feet. Least common are the much larger special-purpose rooms, such as Snake Kiva, which appear to have been used for ceremonies. There are only thirteen of these special-purpose rooms among the Frijoles Canyon cavates, ranging in size from 83 to 189 square feet.

Cavates contain features that tell us much about their inhabitants' activities. Builders often paired plastered or stone-lined hearths or fireboxes with cylindrical smoke holes or vents for air circulation. Many rooms contain plastered recesses and niches carved into the walls and floors in various shapes and sizes, perhaps for storing pots and other household items. Careful examination of the ceilings and walls reveals numerous sockets and holes of different sizes. Some of the large sockets held wooden supports for looms or perhaps upright beams. Smaller holes might have held long, narrow wood rods or sticks on which people hung clothing and blankets. Floor ridges, which are unique to cavates, are mounded seams of adobe running across a

Figure 5.4. The interior of a large cavate with dado, sooted tuff, and petroglyphs.

room, separating lower and higher floor sections. Because the hearth and mealing bins tend to be grouped on the same side of the ridge, it may have divided the room into work and sleeping or storage areas. Earthen metate rests and mealing bins, used for grinding corn, are other built-in features.

Loom anchors offer one of the most interesting clues to the use of some cavates. The only remaining traces of weavers' looms are large sockets in cavate ceilings and rows of small wooden loops set deep into the floors. The ceiling sockets held wooden hooks from which the upper bar of a vertical loom was suspended, and the weaver tied the lower bar to the wooden loops. Loom anchors appear in both habitation and special-purpose cavates but are rare in masonry pueblos such as Tyuonyi (though this may also be the result of poor reporting by early archaeologists). When we find weaving features in the large pueblos, they are usually in kivas. In the cavates, both loom anchors and mealing bins generally appear in second-story or higher cavate rooms. The upper-level rooms may have been good places in which to work, because many of them opened to the outdoors and had natural light. Interior rooms in masonry pueblos and cavates are very dark, and it is difficult to imagine anyone grinding corn or weaving cloth in such a small space by firelight. Indeed, people probably carried

Figure 5.5. Snake Kiva. The broad zigzagging band is the horned serpent's body; his head is on the left.

Figure 5.6. Metates and mealing bins in a Bandelier cavate. Note the scuffmarks left on the wall by grinders' feet.

out many of their daily tasks on the roofs of the masonry rooms fronting the cavates.

Storage rooms were the second most common type of cavate in Frijoles Canyon. Archaeologists call them storage units because they have few or none of the interior features used in daily living. They tend to be smaller than the habitation cavates and are often sooted but not plastered. A few are finished with a plaster dado, which might mean that their function changed over time, from habitation to storage, or vice versa. Storage chambers vary in size from large niches in the walls to individual rooms. Some sit high on the cliff face, reachable only by hand-and-toe holds or by ladder from the roofs of nearby rooms. A few storage cavates in Frijoles Canyon still contain prehistoric corncobs and vessels. We assume that stockpiling food for the future was essential, and these small, featureless cavates served ideally for such use.

The rare special-purpose cavates, larger than living rooms, tended to have a greater number and variety of interior features such as hearths and loom anchors in the floors, as well as painted and incised plaster. Edgar Hewett called these cavates "kivas," but Adolph Bandelier's Cochiti guides during his visit to Frijoles Canyon in 1880 insisted that the cavates were houses and the kivas were in the valley below.

Were these cavates kivas? Unlike most kivas, the cavates have no distinct sipapu, or spirit hole, in the floor. Though large by cavate standards, they are much smaller than kivas found in contemporaneous large pueblos. On the other hand, most of these chambers are physically separate from other cavates, which gives them a certain prominence. Like Snake Kiva, they tend to have wall paintings and incised images in their interiors. Kenneth Chapman, who studied the incised images

Carved in the Cliffs 41

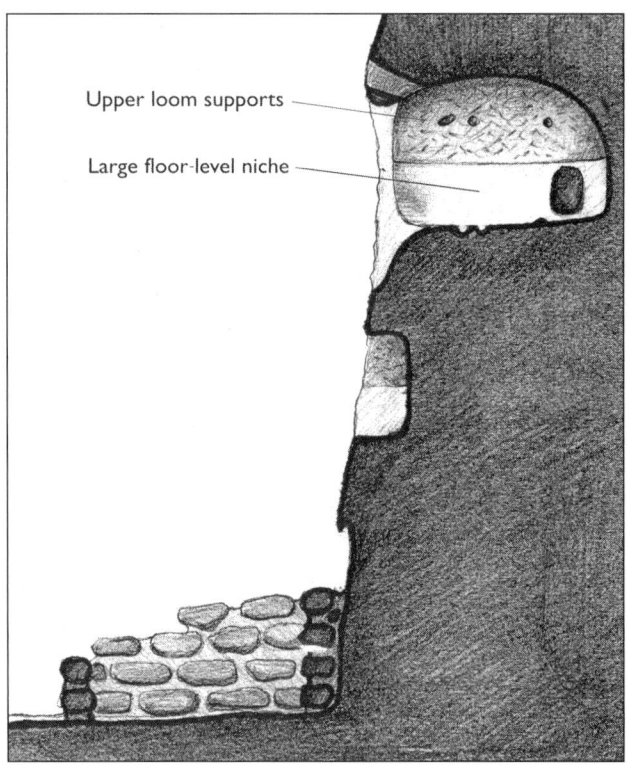

Figure 5.7. Cross section of a cavate room positioned high on a cliff face.

Figure 5.8. A storage cavate containing a clay *puki*, an implement used as a base in molding pottery.

in Frijoles Canyon cavates, thought that these rooms had ceremonial importance because of their paintings of birds, snakes, humanlike figures, and geometric designs, which in Pueblo culture are usually associated with ceremonial practices. Art historian J. J. Brody points out that except for decorated pottery, most kinds of Pueblo art made before 1900 were created as components of rituals. For these reasons, it seems likely that the thirteen largest cavates were used in part for religious activities, although the rituals performed in them might have differed in important ways from those conducted in larger kivas.

With a summary in hand of what is known about cavates, we can return to the question raised at the beginning of the chapter: why did Pueblo people build dwellings in the cliffs? Perhaps the answer is simple: it was, for the most part, easier and more practical to carve and shape cavates than it was to build freestanding masonry pueblos. As far as we can tell, the soft volcanic tuff was fairly workable, making it relatively easy to tailor spaces for dwelling, storage, or special use, and might have required the use of fewer resources such as mud mortar and shaped stones. Furthermore, the rock provides natural insulation in all seasons.

It is clear by now that the cavates were built and used at the same time as the large, freestanding masonry pueblos, but the question remains, how did the two differ? One of the oldest and most compelling arguments is that people used the cavates seasonally, primarily in the winter, but lived in the large pueblos year-round. Those who argue for winter use point to the journals of Adolph Bandelier, who wrote that his companion from Santa Clara Pueblo described the cavates as winter habitations. The presence of small ceremonial rooms and the absence of large kivas suggest indoor, possibly winter ceremonies held by a small number of participants, in contrast to large public ceremonies performed in kivas and plazas during warm summer weather.

Alternatively, the well-insulated cavates stay cool in the summer heat, which makes them practical summer residences as well. The presence of loom anchors and mealing bins in second- and third-story cavate rooms that open to the exterior may mean that weaving and grinding meal were summer activities, too. After spending several seasons working inside the cavates and noticing how cool they are, I am convinced that they were used year-round. Nonetheless, the question of how and why the cavates were used is far from settled. Very likely, archaeologists will still be debating their purpose for many years to come.

**The Cavates Historically and Today**

By the mid-1500s, many of the people who lived in the Frijoles Canyon cavates had moved closer to the Rio Grande. The cavates remained uninhabited until the Pueblo Revolt of 1680–1692, when once again, a few became homes for those seeking refuge from the Spanish. Though Pueblo people no longer used the cavates after 1700, the dwellings remain alive in modern Pueblo memory, perception, and tradition. Pueblo people still visit them and acknowledge them as an integral part of an ancient landscape to which they are strongly connected.

Throughout the 1800s and 1900s, shepherds and hunters used Frijoles Canyon, leaving their mark by carving inscriptions on cavate walls. Mounds of sheep dung covering cavate floors and remnants of wood, metal, and glass debris strewn on the talus slope tell their story. By the turn of the twentieth century, Frijoles Canyon and the cavates were attracting attention for their archaeological value and natural beauty. Beginning in 1910, a distinctly American approach to archaeology was promulgated by Edgar Lee Hewett and his students, including Kenneth Chapman, Jesse Walter Fewkes, Sylvanus Morley, J. P. Harrington, and Jesse Nussbaum, to name a few. Hewett's innovative "experiential learning" included the School of American Archaeology, which became the School of American Research (now the School for Advanced Research), and the Museum of New Mexico. These field schools were focused on excavating and restoring structures such as Cave Kiva, Ceremonial Cave (now Alcove House), Tyuonyi Pueblo, and other larger masonry pueblos of the Pajarito Plateau. During the field schools, some students took up residence in the cavates. Later in the 1930s and 1940s, the cavates and other ancient Pueblo sites in Frijoles Canyon became places for recreation and exploration by participants in the Civilian Conservation Corps and the Manhattan Project.

Since the 1980s, large-scale archaeological studies (Bandelier Archaeological Survey) and systematic documentation of all the Frijoles Canyon cavates through photography, mapping, and laser scanning (Vanishing Treasures Program) have been carried out to better understand cavate construction and use and to protect them as both constructed and natural heritage. How does one preserve these extraordinary dwellings, however, when the landscape is constantly changing and naturally reshaped by increasingly dramatic weather events and erosion? One answer may be in nature itself. New studies in the canyon are focusing on understanding natural preservation processes, such as the formation of thin mineralogical and biological rinds on the tuff surface. The reddish patina and lichens you see on the cliffs may essentially improve the strength and durability of the fragile, glassy ash and protect the cavates and cliff-carved features such as hand-and-toe hold trails and petroglyphs from rapid erosion and loss.

No matter what our specific interests in the cavates, they are fascinating to us all. They are unique for their construction and special features and especially because they illuminate the lives of former inhabitants. It is the cavate that captures in its architectural details the daily activities of grinding corn and weaving fabric. In the cavates, rather than the freestanding pueblos such as Tyuonyi, the details of ancestral Pueblo people's domestic lives are still visible. Archaeologists will continue to study the evidence and grapple with the issues of the cavates' origins and uses; others will be content to peer into the interiors and marvel at what they see.

**Angelyn Bass** is an architectural conservator specializing in the preservation of archaeological and historic sites. Formerly director of the Vanishing Treasures Program at Bandelier National Monument (NPS), she is a research assistant professor in the Department of Anthropology at the University of New Mexico and a principal at Conservation Associates.

Figure 6.1. North wall of Pueblo Bonito.

# Architecture
## The Central Matter of Chaco Canyon

*Stephen H. Lekson*

Architecture is the central matter of Chaco Canyon. Its major ruins make Chaco distinctive, perhaps unique. Our understanding of Chacoan architecture follows directly upon the history of research at this ancestral Pueblo center.

It was the remarkable size of some ruins in Chaco Canyon that first attracted archaeological attention. Chaco's giant "great houses," built of banded masonry and shaped like huge Ds, Es, Os, and Ls, lured the pioneer archaeologist Richard Wetherill away from Mesa Verde. He moved to Chaco, promoted its archaeology, and ultimately led the first major excavations in 1896 at Pueblo Bonito. The work was sponsored by the American Museum of Natural History.

Pueblo Bonito's archaeology was spectacular (plates 8 and 9). The building itself was monumental, and it contained treasures rarely seen in the ancestral Pueblo ("Anasazi") region. Tribal wars broke out between the American Museum, the Museum of New Mexico, the Smithsonian, and other institutions vying for excavation rights at Chaco great houses. Throughout the first half of the twentieth century, the archaeology of Chaco Canyon was fragmented into independent projects at various sites.

Professional rivalries adversely affected the science: excavators sometimes published accounts of their work with little or no reference to other projects. And so each great house was seen as a separate entity: Pueblo Bonito and Chetro Ketl must have been independent towns, much like the Pueblo villages of New Mexico and Arizona. The Pueblo Indians share cultures, ceremonies, economies, and architectural traditions, but each Native town in the Southwest is a separate social unit.

Before the development of tree-ring dating, the great houses were assumed to have been of the "Great Pueblo" period, the thirteenth-century climax of the Four Corners Pueblo culture, which included the cliff dwellings of Mesa Verde, Hovenweep, and many other famous sites. Chacoan great houses were larger and better built than the cliff dwellings. Perhaps Chaco was the climax of the climax? Tree rings, however, revealed that Chaco's great houses were built much earlier than the sites on Mesa Verde. Cliff Palace was built about 1250 to 1280 CE, but construction at Pueblo Bonito began about 850 and ended at 1130.

Chaco's great houses were built more than a century earlier than the Cliff Dwellings of the "Great Pueblo" period. Indeed, great houses—with hundreds of rooms, massive masonry, and geometrically formal ground plans—were built at a time when the typical family house consisted of an earth-covered pit structure and a small, almost ephemeral "pueblo" of five or six rooms. More than 90 percent of Pueblo people, during the Chaco era, lived in these serviceable but unimpressive small houses. The contrast between small and great houses was so remarkable, some archaeologists believed that the two architectural traditions represented two distinct cultures or ethnic groups.

Why were great houses so unlike other Pueblo buildings of Chaco's time? Why were great houses concentrated in Chaco Canyon? Was Chaco more than a valley filled with pueblos?

Figure 6.2. Artist's reconstruction of a typical Pueblo I small house. This design, consisting of an arc of rooms and a ramada with a pit structure in front, was common to ancestral Pueblo dwellings throughout the Four Corners region for many generations. The form is reflected in the earliest construction of Pueblo Bonito.

By 1960, the institutional rivalries surrounding Chaco had faded into footnotes, and it was possible for archaeologists to consider the canyon as a whole. The whole was much more than the sum of its parts. In 1965, Gordon Vivian, Tom Mathews, and Bryant Bannister summarized the archaeology of Chaco. Vivian and Mathews, in *Kin Kletso: A Pueblo III Community in Chaco Canyon*, described three different kinds of architecture: first, a dozen great houses (which they called the "Bonito phase"); second, scores of small houses comparable to regular Pueblo construction of the time (the "Hosta Butte phase," named for a prominent landmark); and, third, a small group of late great houses that reflected Mesa Verde styles (the "McElmo phase," the term used at Mesa Verde for the period just before Cliff Palace). Bannister's *Tree Ring Dating of Archaeological Sites in the Chaco Canyon Region, New Mexico*, synthesized and reevaluated the tree-ring dates known from Chaco and offered the first site-by-site descriptions of all the excavated Chaco ruins. The tree-ring dates showed that the three phases—Bonito, Hosta Butte, and McElmo—were contemporaneous. That is, at about 1100, Chaco had three notably different architectures: great-house, McElmo, and small-house styles. Chaco was a complicated place.

These two studies stood for decades as the definitive summary of Chaco archaeology (and they remain remarkably useful even today). When I began to study Chaco archaeology in the mid-1970s, Vivian, Mathews, and Bannister was the standard reference. At that time, the National Park Service's Chaco Project was several years into a long program of excavations and research. That work reemphasized Chaco's unusual and perhaps unique place in Southwestern prehistory.

In the 1970s, after surveying the entire canyon and producing more accurate maps of the great houses, Alden Hayes, then field director of the Chaco Project, produced the first major reevaluation of Chaco archaeology. Hayes was so impressed with great-house architecture that in his 1981 report, he concluded that overlords or agents from the high civilizations of ancient Mexico had traveled to Chaco and transformed local Pueblo architectural traditions into a new great-house style. Many of Hayes's professional colleagues shared his view. The next task of the Chaco Project was to excavate a great house and test Hayes's theory.

I became part of a new staff hired for this major undertaking. Our duties only began with pick and shovel; in winter, we analyzed artifacts and wrote reports. Assigned the topic of great-house architecture, I revisited Vivian and Mathews's report. And I took advantage of all the new data—Hayes's great-house maps, new tree-ring dates, newly published reports on Pueblo Bonito, Chetro Ketl, and other sites—and the opportunity to spend time in Chaco Canyon examining the ruins. With help from several colleagues, I wrote *Great Pueblo Architecture of Chaco Canyon*, published by the Park Service in

Pueblo Bonito at 850–900 CE had large rooms, tall ceilings, and as many as three stories. An entire Chacoan small house could fit into a single room at Pueblo Bonito. But the masonry generally used for small houses was unsuited to the much larger dimensions of great houses.

An early wall around the rear of Pueblo Bonito, built in the early 900s in the old-fashioned way, began to buckle and fail. In about 1020, Chacoan builders saved the old wall—instead of razing it—by surrounding it with another wall, using a newer masonry style. It worked. At Pueblo Bonito, you can almost see the Chaco builders experimenting with masonry and, over several decades, developing the massive, superbly crafted Chacoan walls that distinguish great houses of the eleventh and twelfth centuries from previous buildings. This, I thought, showed that great houses were a local phenomenon rather than a Mexican import. The Chacoans invented the new masonry styles in order to build bigger buildings.

Figure 6.3. Early Pueblo Bonito wall in foreground, with later core-and-veneer wall behind it.

1984, twenty years after Vivian, Mathews, and Bannister's studies.

My conclusions fell more in line with those of Vivian, Mathews, and Bannister than with those of Alden Hayes, who nevertheless read my "schoolboy" drafts with patience and redirected my wilder fancies with wisdom. I argued that Chacoan architecture developed from regional Pueblo, not Mexican, traditions of form and technique. The earliest building at Pueblo Bonito used standard Pueblo masonry of its time to create radically new forms. Whereas traditional Pueblo dwellings had small rooms, low ceilings, and only a single story,

But why build big buildings in the first place? What were great houses? The masonry was local, but Hayes was right: great houses were astonishingly unlike other ancestral Pueblo buildings of their time, or later. Recall that when great houses were at their peak around 1100, almost everyone else in the Pueblo Southwest was living in small houses of half a dozen rooms and a deep pit structure (sometimes called a "kiva," inaccurately). Pueblo Bonito by 1100 covered almost two and a half acres (about 1 hectare), rose to five stories, and had at least seven hundred rooms and forty "kivas," or pit structures. A precisely sited north-south wall divided Bonito's vast enclosed plaza in two, with one enormous great kiva in each half. A large ponderosa pine in the west plaza must have been watered and encouraged to grow, for it was far out of its natural range; presumably, the tree had powerful symbolic value. Just south of the main building, two rectangular platform mounds, each the size of a basketball court, rose over six feet (2 meters) above a surrounding network of roads, compound walls, and other esoteric architecture. We do not know what structures the Bonitoans might have built atop these massive platforms.

Chetro Ketl, with perhaps six hundred rooms, was comparable in size to Pueblo Bonito. A colonnade faced its plaza. Colonnades were not traditional Pueblo architectural elements—they occur only at Chaco and in Mexico.

Four other great houses at Chaco Canyon were built on this expansive scale: Una Vida, Peñasco Blanco, Pueblo del Arroyo, and Pueblo Alto. Other great houses were smaller, but still mammoth compared with small houses.

Pueblo Bonito and the two other earliest great houses (Peñasco Blanco and Una Vida) began, in the late ninth and early tenth centuries, as monumentally "scaled up" versions of contemporary ancestral Pueblo houses. That is, the plan of early Pueblo Bonito resembled a line of regular small-house family dwellings, side by side, but vastly larger and multistoried. After these beginnings, great-house architecture diverged increasingly from contemporary small sites. By 1130, later great houses, such as Kin Kletso, looked very different indeed from the standard six-rooms-and-a-kiva ancestral Pueblo home.

Tom Windes directed Chaco Project excavations at Pueblo Alto. The Chacoans built this large great house—before excavating it, we assumed that it stood three stories high—on the north mesa, three hundred feet above Pueblo Bonito and Chetro Ketl on the canyon floor. Its view extended from the mountains behind Albuquerque to the peaks behind Mesa Verde. Windes discovered that Pueblo Alto was only one story high (though ceiling heights were impressively tall, some more than twelve feet), and despite its 135 rooms, only a few families had lived there. He based the latter conclusion on the absence of features such as fire boxes and storage bins, the marks of everyday domestic use, in most of the rooms at the site. Extending his analysis to Pueblo Bonito and Chetro Ketl, he concluded that few of the great houses, if any, were "pueblos"—that is, they were not towns filled with families.

I reached similar conclusions from another line of evidence: pit structures. Circular, underground rooms at both great houses and small houses are conventionally called "kivas," after the ceremonial structures we see today in the plazas of Pueblo Indian villages. Rio Grande pueblos typically have one or two kivas apiece, continuing the pattern seen at scores of ruins from the fourteenth century onward. Rio Grande kivas are probably modern versions of the great kivas at Chaco, which usually had one or two great kivas per great-house community. Great kivas were there from the beginning, and great kivas are still there today in the plazas of modern pueblos.

What, then, are all the little "kivas" at Chaco and other ancient sites? The scores of small, circular rooms at Chaco and Mesa Verde probably were not kivas like those at modern pueblos, but instead the final and most elaborate form of the pit house. Pit houses—circular or square underground rooms—were the primary residential structures of ancestral Pueblo peoples for at least five centuries before Chaco. They were easy to build, cool in the summer, and warm in the cold desert winters. After about 900 CE, people began to add small aboveground structures behind the pit houses; eventually, those small sites evolved into the pueblos of today. But at modern Rio Grande pueblos, there are hundreds of rooms for each kiva. Recall that, at Chaco's time, a typical small house consisted of only five or six rooms and a pit structure ("kiva"). At Yellow Jacket, the largest Mesa Verde pueblo, that ratio was even lower: three or four rooms per pit structure. With almost two hundred "kivas," Yellow Jacket has been interpreted as a ceremonial center, but the same interpretation could be applied uniformly across the Chaco and Great Pueblo periods: almost every large site had many "kivas" compared with relatively few rooms.

If the small, circular pit structures were actually houses (or rather, parts of small houses), then we can estimate the number of families living in a great house by its number of original "kivas." (Later

Figure 6.4. View of the southeast section of Pueblo Bonito, showing a series of circular and rectangular rooms.

reoccupations of great houses often added more "kivas," or pit houses). At Pueblo Alto, there were six or seven original "kivas," and this number supports Windes's conclusion: few families lived in Pueblo Alto or other great houses.

The low room-to-"kiva" ratio is reversed at some great houses, where a circular pit structure might be associated with thirty or more rooms. Families living in great houses had many more rooms and much more floor area than did other ancestral Pueblo families. The great houses' hundreds of extra rooms and thousands of square feet represented something more than domestic households. Analyses by Jill Neitzel suggest that Pueblo Bonito was an elite residence—that is, the home of a small group of families who were politically, socially, or ceremonially more powerful than other people in the region and who demonstrated their importance by living in very large houses. They lived in large, elaborate pit houses with suites or apartments of perhaps a dozen rooms attached—rather like a traditional small house, but much bigger.

Housing was only a small portion of each great house. Scores of other, nonresidential rooms probably served a variety of purposes: storage (approaching warehouse dimensions), chambers reserved for ceremonial or political functions, temporary housing for visitors or laborers. Elsewhere, I have referred to the Chacoan great houses as "palaces," a word that reflects their multiple functions as elite residences, warehouses, administrative offices, and ceremonial centers. Indeed, Chacoan great houses are comparable in size and complexity to the famous Minoan palaces of Crete. This is not to say that Chacoan society rivaled or even resembled that of ancient Crete; great houses were very modest palaces in almost every regard. But Chaco great houses served the same functions as palaces in Crete and Mesoamerica.

In the early 1980s, I attempted to estimate the amount of labor required to build great houses, from digging mortar and shaping stones to building walls. I concluded that great-house construction required about forty thousand person hours of building labor every year between 1050 and 1125. This must have come not from the great-house residents but from outside. Those who lived in the buildings organized and planned but did not labor.

Some great-house building may have been principally for the purpose of massing. "Massing" is an

Figure 6.5. Large, deep nonresidential rooms in Pueblo Bonito may have been used for bulk storage.

architectural term that the archaeologist John Stein and others employ to suggest that the Chacoans built great houses to create impressive forms. At Pueblo Bonito, for example, most of the building consisted of dark interior spaces, deep below multiple upper stories and far removed from plaza-front doorways. Some of these rooms could have been used to store all sorts of objects, as Pueblo people do today, but it seems likely that the creation of impressive, monumental forms was also a motivation.

The McElmo phase, late in Chaco's history, is a case in point. John Stein and Ruth Van Dyke have suggested that the compact McElmo structures of the early 1100s represent pure massing—that is, building to build, building to impress. With their small, compact rooms and multiple stories, McElmo structures stick up almost like towers. They show little evidence of use; most, in fact, have only one or two "kivas" (presumably indicating only one or two families in residence) and no trash middens. The final spurt of Chacoan building created mass as monuments on the Chacoan landscape.

The large, geometrically formal great houses were meant to be seen and to create an impression from the cliffs above Pueblo Bonito and Chetro Ketl and from the "roads" on the valley bottom. Their layout demonstrates canons or rules in construction and form, just as medieval cathedrals or Buddhist stupas have common features, ground plans, and geometries. Anna Sofaer, John Stein, and other researchers have begun to unravel those rules.

After a hundred years of scientific research in Chaco, we recognize that it was more than a canyon full of pueblos. Chaco Canyon was an architectural composition, combining a dozen great houses and scores of other monumental features with a distinctive, even unique landscape to create a small planned city. When I introduced the term "downtown Chaco" (plate 15) in 1981 for the complex of buildings around Pueblo Bonito, Chetro Ketl, and Pueblo Alto, I was not the first to suggest that Chaco was formally designed. In 1978, John Fritz demonstrated that downtown Chaco reflected two major axes: a perfect north-south meridian from

Pueblo Alto on the north to Tsin Kletsin on the south and an east-west axis from Pueblo Bonito to Chetro Ketl. Gwinn Vivian and later Tom Windes carefully mapped the dense nexus of roads and stairways that connected Pueblo Alto, Pueblo Bonito, and Chetro Ketl, and John Stein and his colleagues extended that constructed landscape to encompass all of downtown Chaco and beyond. Stein, working with Richard Friedman, Taft Blackhorse, and others, now posits a sculpted Chaco with great houses, complexes of pyramid-like platforms, roads, broad ramps and stairways, massive walls enclosing whole districts of buildings, and an astonishing variety of other monumental features. Interspersed among all these structures was a complex series of irrigation systems and gardens. The built environment of Chaco resembles, in conception if not in scale, the ceremonial centers of ancient Mexico and the Maya region.

Stein and his colleagues and Anna Sofaer and her Solstice Project have explored the layout of downtown Chaco, around Pueblo Bonito and Chetro Ketl, showing that the whole area was almost theaterlike as a setting for pomp and circumstance, ritual and procession. The flat, featureless ground between Pueblo Bonito and Chetro Ketl may have been a central stage for ritual; Stein and his colleagues suggest that the canyon walls there had distinctive acoustical properties that magnified the sounds of ceremony. The canyon's rim provided unmatched viewing for hundreds, even thousands, of observers, perhaps pilgrims reaching the canyon on the ceremonial roads. The terraced roofs of the great houses themselves were perhaps VIP seating for hundreds of elites. Downtown Chaco was a city, but it was also a sacred theater, part natural and part carefully constructed.

Chacoan monumental building traditions extended well beyond the canyon. Great houses, usually much smaller than those in the canyon but built with similar technologies and according to similar design canons, are found as far as 150 miles away. These sites are quite variable, but the range of forms and masonry styles seen in so-called outliers is met or exceeded by architectural variation within Chaco Canyon great houses, which range from very large to rather small buildings. Roads, earthworks, and other major elements of Chacoan building appear throughout the region. Together, shared forms and details suggest a regionwide architectural tradition. Clearly, that tradition found its most remarkable expression in Chaco Canyon itself, and it seems likely that Chacoan architecture outside the canyon was inspired by or referred back to Chaco Canyon. Architecturally, Chaco Canyon was the center of the larger region.

Alden Hayes and Tom Windes's discovery of a sophisticated line-of-sight communication network extends the built environment of Chaco Canyon to its region, perhaps to its very farthest corners. This line-of-sight system, presumably operated with smoke or mirrors, paralleled the more famous Chacoan roads and ultimately may prove to be more important to understanding Chaco and its world. Strategic high points throughout the Chaco region feature large fire boxes and shrines that are visible from more distant high points where outlier great houses are situated. For example, Katy Freeman, a high school student, discovered that Chimney Rock Pueblo was positioned to allow line-of-sight to Huérfano Mountain, in northern New Mexico. Huérfano Mountain (which has the remnants of many fire boxes and shrines) in turn has direct line-of-sight to Pueblo Alto. Messages could have been passed from Chimney Rock, at the northeastern edge of the Chacoan region, to downtown Chaco in a matter of minutes—if people were manning the "repeater" station at Huérfano Mountain. The signaling system would have required careful management, staffing, and coordination. Its existence throws light on the position of Pueblo Alto, with its remarkable views to the north; Peñasco Blanco, with its views to the west; and probably most other Chaco Canyon great houses—field of view was a critical factor in placement.

For Chacoan architecture, we must understand far more than masonry details and ground plans of individual great houses. This was a monumental tradition, designed and built on citywide and even regional scales. Its shape reflects social, political, and ceremonial dimensions. In concept and execution, Chaco dwarfed other Southwestern architecture, with the possible exceptions of the massive ceremonial complex at Aztec Ruins (which flourished in

Figure 6.6. New Alto, above Chaco Canyon on the north side, is a late great house with a dominant view of the surrounding countryside.

the twelfth and thirteenth centuries) and the great city of Paquimé (fourteenth and fifteenth centuries) far to the south.

Andrew Fowler, John Stein, Keith Kintigh, and others have shown that after construction ceased in Chaco Canyon about 1125, the great-house tradition survived for decades at sites throughout the old Chaco region. As I explained in my book *The Chaco Meridian* (1999), Aztec and perhaps Paquimé were direct architectural heirs of Chaco. In any event, Chacoan architecture deeply influenced subsequent Pueblo building and history. Chaco great houses were, indeed, the first structures in what we today call "Pueblo style": terraced, multistory architecture. Modern Pueblo style replicates the forms of Chacoan buildings but probably not the social, political, and ceremonial complexities that shaped them.

Chacoan architecture was distinctive, but was Chaco unique? Like any other great tradition, Chaco had its particular history. But we cannot lose sight of its context: arable North America in Chaco's time was densely populated by monumental social-ceremonial centers, some smaller but most larger than Chaco. The Chacoans clearly knew the polities of Mexico far to the south; closer to home, however, were the numerous "chiefdoms" of the Mississippi Valley, with their huge ceremonial and political centers.

Cahokia, the largest Mississippian site and contemporary with Chaco, provides an interesting comparison. Located near St. Louis in the rich farmlands of the American Bottom, Cahokia boasts an earthen pyramid as large as the largest in Mexico. Scores of smaller "mounds" and monuments define a complex ceremonial cityscape that was comparable in spatial extent to Chaco Canyon but larger in population. No more than three thousand people lived at Chaco, in comparison with ten thousand or more at Cahokia.

Thus, Chacoan architecture was unique within the Southwest of its time, but it was comparable to the architecture of hundreds of other monumental centers of the eleventh and twelfth centuries in Mexico and North America. We rightly marvel at Chacoan building, but we should not be unduly surprised that Native Americans built large,

Figure 6.7. Aztec West, a major Chacoan great house along the Animas River. Its great kiva was reconstructed.

monumental ceremonial-political centers in the ancient Southwest: such centers were present almost everywhere else that corn could be grown on Chaco's continent.

Of course, most of those centers, such as Cahokia, were built in far better farming environments than Chaco's. It is easier to imagine ten thousand people living and farming in the American Bottom and building Cahokia than to envision three thousand surviving at Chaco, much less producing sufficient corn, beans, and squash to support the influx of labor for building great houses, platform mounds, roads, and other monuments.

Chaco was monumental. *Modestly* monumental —Chaco was not Cahokia, nor was it Tula. Still, it was far more than a valley full of pueblos. Chacoan architecture itself is our best and biggest clue that Chaco was actually a city, with a resident elite controlling a region. Great houses themselves represent stratified housing. They "cost" a lot to build per square foot of floor space, and their builders used superior materials and construction techniques. They were built to last. A small site—the normal family home—might be abandoned or rebuilt every generation, but Pueblo Bonito stood for nearly three centuries. Large sections still stand today. Great houses strongly suggest that their residents were of a different class from the people who lived in small houses, a fact born out by many other lines of evidence related in the chapters of Noble's *In Search of Chaco*.

**Stephen H. Lekson**, professor of anthropology and curator of archaeology at the Museum of Natural History, University of Colorado, Boulder, has conducted Chaco-related research since the mid-1970s. He was a member of the National Park Service's Chaco Project and director of its later Chaco Synthesis. His books include *A History of the Ancient Southwest* (SAR Press 2009) and *The Archaeology of Chaco Canyon* (SAR Press 2006).

Figure 7.1. Aerial view of Wupatki Pueblo, with its ballcourt (foreground) and "amphitheater" (the circular unroofed structure).

Plate 1. Artist's conception of Little Box Canyon Pueblo in Wupatki National Monument as it might have looked in late summer about 1190 CE.

Plate 2. Tule duck decoy from Lovelock Cave, Nevada.

Plate 3. Desert pond on sagebrush flat east of Denio, Nevada, with Bilk Creek Mountains in background, Humboldt County, Nevada.

Plate 4. Barrier Canyon–style paintings with dot decorations and supernatural snakes, Clear Water Canyon, eastern Utah.

Plate 5. Artist's reconstruction of Sand Canyon Pueblo around 1250.

Plate 6. Pottery time line for the Sierra Sin Agua, roughly 550 to 1500 CE. The kinds of vessels shown were all part of the Kayenta ancestral Pueblo and Hopi ceramic traditions, centered just to the east and north of the Sierra Sin Agua.

Plate 7. Wukoki Pueblo in Wupatki National Monument.

Plate 8. Kivas in morning fog at Pueblo Bonito. The great house includes thirty-three kivas.

Plate 9. Craft arts from Pueblo Bonito excavated by the Hyde Exploring Expedition of 1896–1899 and now in the American Museum of Natural History. *Left:* reconstructed turquoise-encrusted cylinder; *center:* deer bone spatula or scraper inlaid with jet and turquoise; *right:* McElmo Black-on-white pitcher.

Plate 10. Following Tohono O'odham tradition, Juanita Ahil gathers ripe saguaro fruits in early summer in the 1960s.

Plate 11. Indigenous corn (*Zea mays* L.) in the southwestern United States and northern Mexico today includes varieties that vary notably in ear size, kernel color, and kernel texture (flour, flint, pop, sweet, dent). The ears shown here represent (a) Santo Domingo blue flour, (b) Isleta white flour, (c) Hopi red flint, (d) Hopi red flour, (e) Mojave white flour, (f) Mexican Harinosa de Ocho white flour, (g) Cochiti purple flour, (h) Acoma yellow flint, (i) Tohono O'odham yellow flour, (j–k) Chapalote brown pop/flint, (l) Mexican pink pop, (m–n) Hopi blue flint and flour, (o) Hopi red-striped flour, (p) Acoma white pop, and (q) Tesuque white sweet. Such diversity in corn has been manicured for both ceremonial and culinary needs.

Plate 12. Style III bowl showing man and woman handling parrots, perhaps as part of a ceremony. The woman has a bird head, which may represent a mask, wears a string apron, and appears to have sandals on her feet. The man wears a headdress and a belt. Each uses a hoop or perch for a bird and carries a crooked stick. A bow and two arrows and a burden basket with a third parrot rest on the ground.

Plate 13. Fremont ceramic pitcher with coffee-bean appliqué, from the Round Springs site, central Utah.

Plate 14. A small cliff house in the Grand Gulch area of southeastern Utah.

Plate 15. View of the central-western portion of Chaco Canyon, with Pueblo Bonito at bottom center and Pueblo del Arroyo immediately above it along Chaco Wash.

Plate 16. View of Frijoles Canyon, Bandelier.

Plate 17. One of numerous ancient trails on Tsankawi Mesa.

Plate 18 (*right*). Paleoindian projectile points from the Sierra Sin Agua. *Top row, three left-most*: Clovis points; others are later styles.

Plate 19 (*bottom left*). An atlatl (*right*) and stone-tipped foreshafts for darts, from Sand Dune Cave, Utah. Although they date from the first century CE, they illustrate important parts of the hunting technology used by earlier Paleoindian and Archaic peoples of the Sierra Sin Agua.

Plate 20 (*bottom right*). Split-twig animal figurines, probably representing deer or antelope, from Walnut Canyon. Such objects may have been used during late Archaic times as symbols of group identity or as totems in rituals promoting hunting success.

Plate 21. Mummy Cave, Canyon del Muerto.

Plate 22. Paintings in the Barrier Canyon style at Sego Canyon.

Plate 23. Wupatki Pueblo.

Plate 24. Chaco Wash in a summer flood.

Plate 25. Aerial view of Sunset Crater.

# Wupatki Pueblo

## Red House in Black Sand

*Christian E. Downum, Ellen Brennan, and James P. Holmlund*

The Wupatki Basin makes a picturesque but seemingly improbable spot for a major center of Hisat'sinom life. Towering cliffs of black basalt rim a red sandstone desert where soil lies thin atop bedrock, trees are stunted and sparse, and even drought-adapted shrubs and grasses struggle to survive. Hot and dry for much of the summer, the basin averages a little over eight inches of precipitation a year—less than the quintessential desert city of Tucson, Arizona. In winter, fierce blizzards can drop more than a foot of snow and plunge the temperature to below zero. Springtime brings howling winds to whip up clouds of reddish brown dust and rearrange the thick, rippled dunes of black volcanic cinders.

Yet, in this unlikely setting, the walls of Wupatki Pueblo began to rise nearly nine hundred years ago. Within about three generations, it grew into a four-story, hundred-room structure and became the heart of a thriving pueblo community. At Wupatki's peak in the late 1100s, it stood as the region's largest and tallest pueblo—surely a cultural center for people of the surrounding region. It likely served as a gathering place, a trading center, a treasury of exotic goods, a landmark, and a place for sacred rituals and ceremonies.

The name Wupatki can refer not only to the pueblo but also to Wupatki National Monument, the Wupatki Basin, the Wupatki Archaeological District (a property on the National Register of

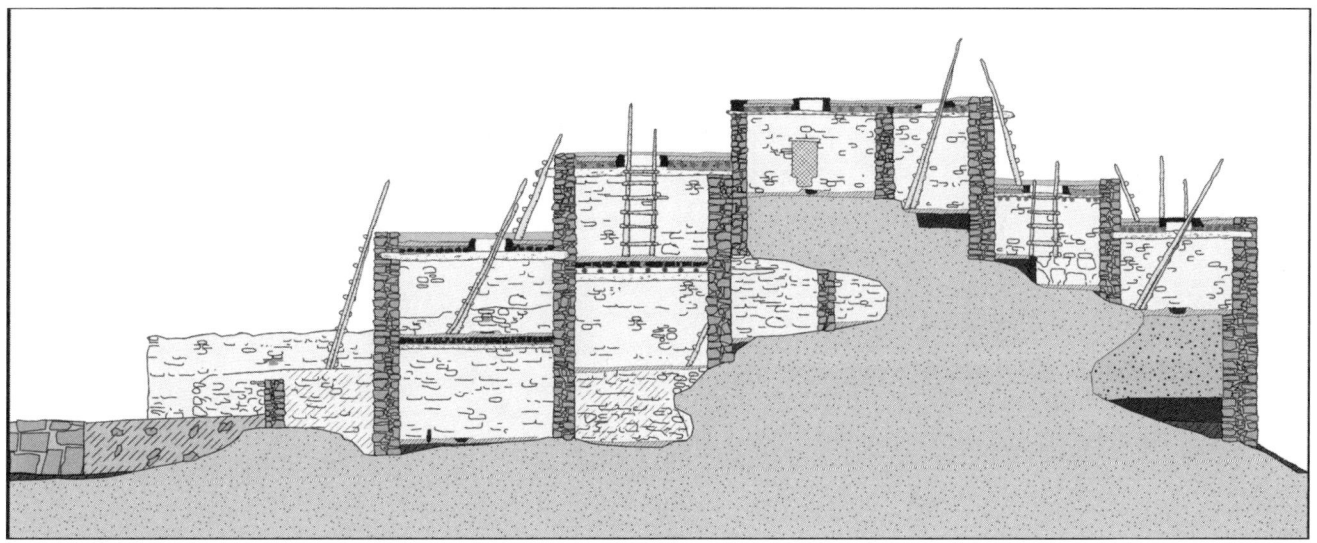

Figure 7.2. Schematic profile of the south room block of Wupatki Pueblo, reconstructed from archaeological evidence gained during excavations in 1933 and 1934.

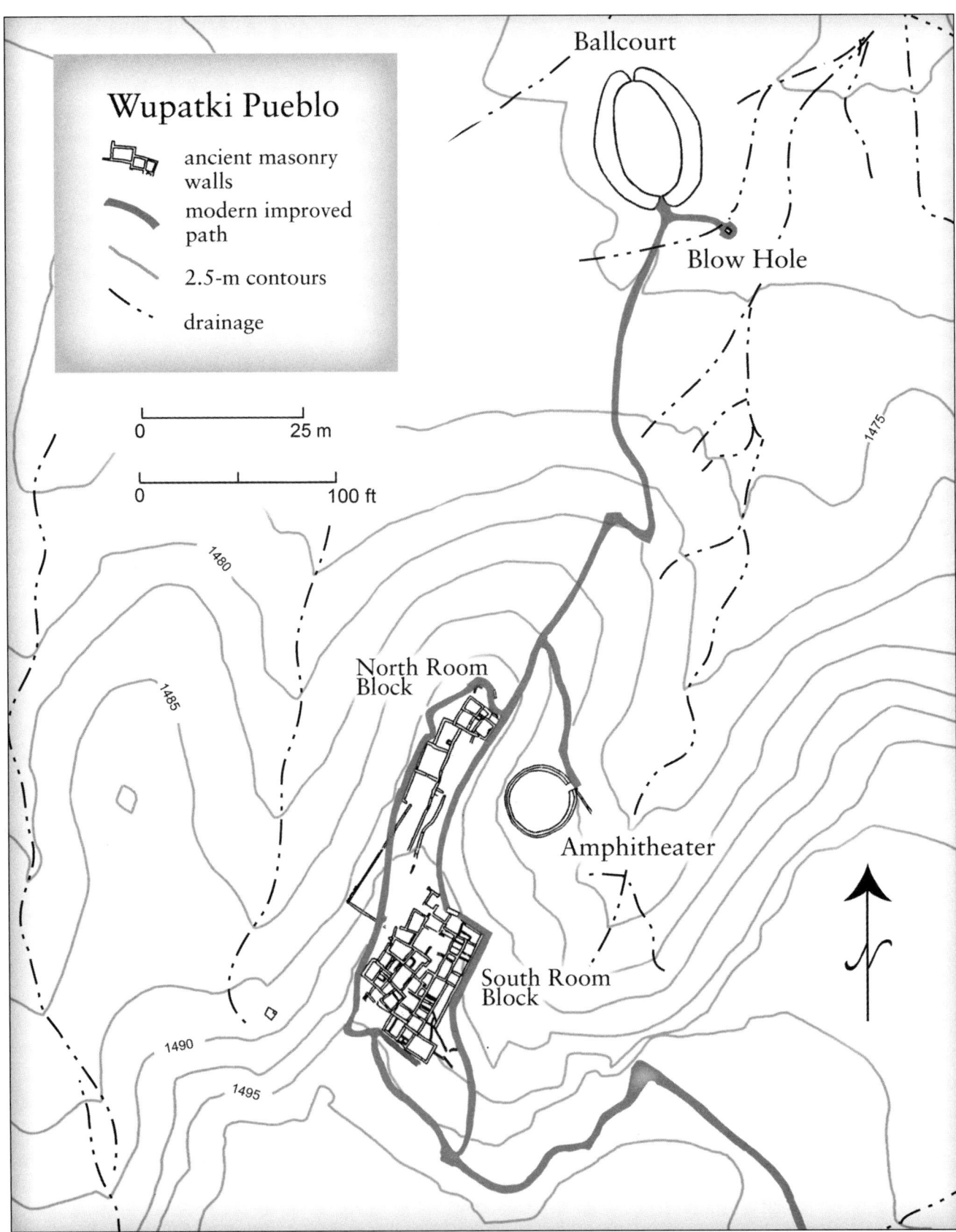

Figure 7.3. Plan of Wupatki Pueblo.

Figure 7.4. Aerial view of the north and south room blocks at Wupatki.

Historic Places), and the ancient pottery type Wupatki Black-on-white. The word has traditionally been thought to derive from the Hopi name Wupakikuh, "Tall House Ruins." Wupatki Pueblo (plate 23) received that name officially, probably at the suggestion of archaeologist Harold Colton, when President Calvin Coolidge set aside land in 1924 to create Wupatki National Monument. Another Hopi name for Wupatki, preferred by some Hopi elders and scholars, is Nuva'ovi, or "Place of the Snows."

Wupatki Pueblo emerged in the early 1100s, a time when great Southwestern cultural traditions that had endured for centuries were changing rapidly. To the east, the people living in Chaco Canyon, their influence waning, had mostly stopped building their "great house" pueblos and kivas. To the south, an important phase in the Hohokam way of life was also drawing to a close. Hohokam people, who had for so long expressed their worldview and religion through games played in ballcourts and fiery death rites involving cremation, were just starting a "Classic" period that repudiated such things.

Closer to home, residents of the Colorado Plateau were rearranging themselves on the landscape. From the late 1000s through the 1130s, an enigmatic group of Hisat'sinom known to archaeologists as the Cohonina left their homeland west of the San Francisco Peaks and moved eastward onto the pine- and juniper-clad mesas just south of Wupatki. Around 1140, Hopi ancestors known archaeologically as Kayenta Puebloans began retreating from distant outposts and concentrating in a much smaller territory east and north of the Little Colorado River. After the Sunset Crater eruption, many so-called Sinagua people relocated from the ponderosa pine forests around modern Flagstaff to lower-elevation grasslands to the north and east. At the time of these movements, people in the region were living in small, widely spaced hamlets and villages of scattered pit houses and modest, one-story pueblos.

Only a few decades before Wupatki began to grow, the Sunset Crater volcano (plate 25) had belched its final clouds of steam, leaving the landscape utterly altered. Area farmers, who previously had largely written off Wupatki as too hot and dry for cultiva-ting, probably took note of the thin blanket of cinders covering the earth downwind and down-slope from the volcano. This "black sand" was having a beneficial mulching effect and enabling trees, shrubs, and grasses to thrive. Even if the Wupatki Basin was still not a great place to farm, it might now be among the least bad during bad times. The stage was set for a population boom.

Like many other ancient pueblos, Wupatki developed on the remains of earlier structures. Unfortunately, we know little about them; the evidence consists only of thin deposits of artifacts under rock overhangs, masses of charcoal representing burned and collapsed rooms or pit structures, and a few storage pits.

Once people started to build the pueblo of Wupatki, it grew irregularly. Builders added rooms as they were needed, not aiming for consistency in size, shape, construction techniques, or internal features. Generally, the pueblo grew outward and upward from a core of early rooms placed against and atop the bedrock at the southern end of the village and from a nucleus of rooms built around large boulders at the north end. Eventually, builders joined the north and south room blocks into a single, continuous structure. Interestingly, the earliest rooms at Wupatki display what some researchers have interpreted as a Chacoan style of masonry, with carefully shaped and arranged stones. The roughly hewn, irregularly stacked masonry of the later rooms better resembles that of nearby Pueblo structures to the north and east.

Archaeologists are fortunate to have a large number of tree-ring samples with which to date Wupatki Pueblo. We estimate that the pueblo's builders cut at least 1,670 pieces of wood for roof beams, upright posts, and doorway lintels. Of these, 183 survived to provide tree-ring dates, 90 of them "cutting" dates—that is, the precise year when a tree died, usually because someone cut it down. Nearly half this wood came from Douglas fir and spruce trees that Wupatki's builders could have found only near the San Francisco Peaks, some twelve to twenty miles south of Wupatki.

Tree rings tell us that the first wood-cutting episode took place in the mid- to late 1130s. Additional clusters of dates then appear every seven to fifteen years, perhaps as families grew or

Figure 7.5. View into the Wupatki ballcourt from its narrow northern entrance.

immigrants moved in. After 1200, construction diminished, and by about 1215, Wupatki's residents seem to have stopped building new rooms. A few tree-ring dates from the mid-1200s appear after a gap of a few decades, marking a later reoccupation of the pueblo or perhaps a handful of stubbornly persistent families who had hung on all along.

Probably early during the construction, Wupatki's builders added two large, unroofed ceremonial structures, likely for performances that were accessible to most or all of the pueblo's residents. One, sited about four hundred feet beyond the north end of the pueblo, is an oval depression lined with stone masonry and measuring about forty-five by ninety feet, with narrow doorways at each end. It so resembles Hohokam ballcourts that archaeologists have named it the Wupatki ballcourt. We believe that this structure, along with about twelve similar courts throughout the Sierra Sin Agua, once hosted a game in which players used a rubber ball imported from Mexico. Some Hopis disagree, believing that it more likely served as a dance court, perhaps for rituals involving the Snake Clan.

The other ceremonial structure is a carefully made, circular, stone-walled enclosure just east of the pueblo. Known to archaeologists as the Wupatki amphitheater, the building measures about forty-five feet in diameter and has an encircling bench. It vaguely resembles Chaco-style great kivas or, more accurately, some of the large, kivalike structures built in the northern Southwest after most people had departed Chaco Canyon. We do not know exactly when the ballcourt and the amphitheater were built; neither has produced a tree-ring date. Judging from pottery, both seem to have been used throughout the life of the pueblo.

Another great open-air construction was a large, rectangular plaza built along the west side of the pueblo, now mostly collapsed and eroded. The plaza's builders first erected an impressive retaining wall, in some places up to eight feet high, and then filled the area behind it with stones, trash, and soil. The resulting platform served as a place where people could gather, socialize, work, and perform public ceremonies. Another possible plaza lies along the east side of the pueblo, surrounding the amphitheater. Such large, open-air plazas are rare at twelfth-century pueblos in the region. The two examples at Wupatki, along with the amphitheater and ballcourt, suggest that this pueblo became an important meeting place, with no contemporary equal nearby.

Adjacent to the ballcourt sits an impressive geological feature known as a "blow hole," one of many in the area. At the surface, blow holes look simply like small cracks or holes in the bedrock, but they lead deep underground to a massive set of interconnected, natural rock chambers. Depending on atmospheric pressure, a blow hole either expels or inhales air at a high velocity. On a hot summer day, cool, musty air rushes out of the Wupatki blow hole with great speed and force. In winter, wind from the blow hole feels relatively warm against the cold surrounding air.

Ancient people might have interpreted blow holes as meaningful links between the underworld, the surface world, and the sky. Modern Pueblo people commemorate the connection between surface and subsurface worlds with the *sipapu*, a small hole in the floor of a kiva. To the Hopis this feature symbolizes the original place of emergence, the Sipapuni, where the first humans entered this world by climbing up from the underworld.

The people of Wupatki probably came from many places at different times. In archaeological terms, the "cultural affiliation" of Wupatki's residents remains a puzzle. Anthropologists have conducted few detailed studies of burials at Wupatki, but the little information that exists portrays a diverse population. Some people had flattened skulls, a slight deformity caused by harmless pressure exerted by cradles on the growing skull bones of infants. This pattern was common among Sinagua people to the south. But several people buried at Wupatki had undeformed skulls. Some people were placed in the grave in a "flexed" position, with the body tightly folded; others were laid out fully extended. At the time, flexed burials were common among people mostly to the north, and extended burials were more frequent in the south. Wupatki has even yielded one cremation burial, a practice usually associated with the Hohokam.

Architecture and artifacts tell a similarly complex story. Architectural styles at Wupatki show both Chacoan influence and, later, similarity to nearby Kayenta pueblos. The pueblo's ballcourt and an abundance of shell jewelry suggest that its people also had some affiliation with Hohokam people—or perhaps Hohokam ideas—from far to the south. Much of the pottery excavated at Wupatki originated in the vicinity of modern Flagstaff, but a substantial portion came from the Kayenta country to the north. Remnants of cotton cloth show styles and decorative techniques found in several parts of the Southwest.

Consistent with these archaeological findings, Hopi and Zuni oral traditions portray Wupatki as a place where people of diverse origins lived briefly on their way to their ultimate stopping points, villages on the Hopi Mesas in Arizona and in the vicinity of modern Zuni Pueblo in northwestern New Mexico. The Havasupais, who now live along the south side of the Grand Canyon, have their own oral traditions linking them to Wupatki. Navajo people, too, hold strong ties with the Wupatki area, having lived there since at least the early 1800s.

Archaeologists have debated over Wupatki for more than a century, ever since the Smithsonian Institution's Jesse Walter Fewkes interviewed Hopis in the 1890s and early 1900s, trying to piece together the routes Hopi clans had followed during their migrations toward the Hopi Mesas. In the 1990s, David Wilcox, of the Museum of Northern Arizona, interpreted Wupatki as an outgrowth of the Chaco cultural system. To Wilcox, Wupatki was a regional political and trade "nexus," strategically situated between desert scrublands and pine forests, connecting peoples to the south with the Kayenta world to the north and the Chacoan world to the east. Wupatki, a local form of a Chacoan great house, arose because of the ambitions of people who traded—literally—on the mystique of Chaco Canyon.

More recently, Glenn Stone and Chris Downum

Figure 7.6. Museum of Northern Arizona archaeologists excavating Wupatki Pueblo in the early 1930s.

proposed that Wupatki indeed grew because of its role in political processes, but those processes were largely local. In this explanation, intense competition over farmland in a risky agricultural environment led to violent conflicts among groups of different cultural origins. Some groups organized themselves along ethnic lines to protect and expand their landholdings. Pueblos such as Wupatki, the Citadel, and other large, visually prominent communities served as symbolic expressions of political power and as central places in local political, religious, and economic life (plates 1 and 7).

In the view of some Hopi scholars, Wupatki came into existence and fulfilled its destiny as intended by spiritual forces. In this view, no more needs to be asked of the pueblo, and no scientific explanation is required. Its true meaning is to be found in its place in Hopi cultural history and its significance to Hopi people.

Tree-ring evidence shows that by the mid-1200s, Wupatki Pueblo lay empty. Whatever sparked the exodus, it surely was wrenching. The pueblo represented an enormous investment of human labor, emotion, and hope. The departure, though, seems to have been orderly. Unlike many other ancient Southwestern pueblos, especially those of later time periods, Wupatki was not burned. We think that the intact roofs, sealed doorways, and many valuable things left in the rooms imply that at least some people expected to return someday.

Why did they decide to leave? A severe drought between 1215 and 1221 correlates neatly with the cessation of building. Studies of plant remains from pack rat nests built up over many years in Wupatki National Monument indicate that by the early 1200s, the area had been largely denuded of trees and shrubs, which people needed to heat their homes and cook food. Farmers had cultivated nearly every possible location that could have been planted, and with repeated use, the fields lost nutrients vital for sustaining crops. Perhaps new pueblos, new communities, and new opportunities promised more than Wupatki could offer to its younger and more restless residents. Internal dissension or threats from neighbors, as well as prophesies or other beliefs, might also have influenced decisions to leave.

For a brief time after Wupatki fell silent, life there survived in the memories of people who had resided in or visited the great place. Eventually, those memories passed away, too. Wupatki Pueblo receded into legend, becoming part of the stories elders told as they recounted the many places where their ancestors had lived and the adventures they had experienced while walking the tangled paths of ancient migrations. Today, under the care of the US National Park Service, Wupatki enables the footprints of the present to mingle with those of the past in the fine red dust of ancient floors.

**Christian E. Downum** is a professor of anthropology in the Department of Anthropology at Northern Arizona University and former director of the NAU Anthropology Laboratories. He has conducted archaeological research in the Sierra Sin Agua since 1982, mostly at US national parks and monuments. He also serves as archaeological advisor to the Footprints of the Ancestors project, an intergenerational learning program that teaches Native American youths about the ancient places of the American Southwest.

**Ellen Brennan** is the cultural resource program manager for Grand Canyon National Park. She holds a master's degree from Northern Arizona University, and her interests include ceramic chronologies, architecture, and Grand Canyon archaeology.

**James P. Holmlund**, president of Western Mapping Company, in Tucson, Arizona, has created hundreds of precise maps of ancient pueblo and pit house sites in the Southwest, including the first high-precision GPS map of Wupatki Pueblo. Recently, he pioneered the use of three-dimensional lidar scanning for documenting artifacts, architecture, and rock art.

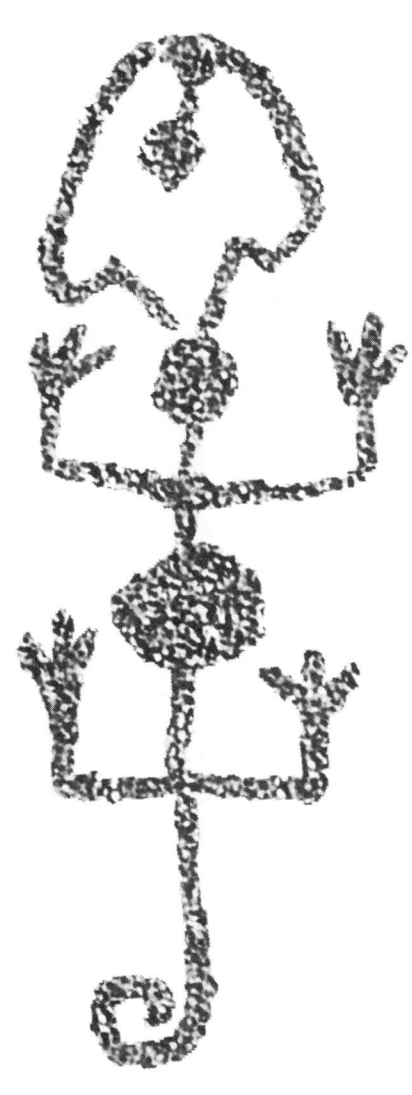

Figure 8.1. Untitled drawing by Kia Gaspar, 1978, perhaps illustrating a harvest festival.

# Zuni Religion and Philosophy

*Edmund J. Ladd*

All of Zuni life, from birth until death, revolves around rituals, ceremonies, special observations, and the crises of transition from one age or status to another. The Zuni believe that everyone carries within himself his* own personal "life road" (*/onnane*). The lifeway, also known as the breathway, the path, or the road, is "kept" by those spirit beings who watch over our roads (*ho/n/a:wona:willanp/ ona*). Personal conduct in this life ensures a smooth and long road—that is, a long and healthful life. It is often said by the elders of the tribe, "There is but one joyous life—you should love each other." This enjoyment of the one life is achieved through active participation in and observation of the various rituals and ceremonies provided by the Zuni socio-religious system to supplicate and appease the spirit beings who keep our roads.

A child is born into the system. His position within the kin-clan group is established at birth. What he is called, how he is called, which of the six men's societies (kivas) he will belong to, and who he can or cannot marry are established. Position at birth determines both his future behavior and how others will behave towards the child. He belongs to his mother's clan, a connection that immediately establishes certain behavior patterns and also, to some degree, determines which position of responsibility he will hold in the religious system. He is a "child of" his father's clan, whose members will provide support in various religious and life crises throughout his life.

Thus, the individual knows from childhood his position in the system. He has a keen awareness of the total religious system and the ceremonial cycles and an especially acute awareness of those activities in which participation is required. Every kiva member knows, for example, when it is time for his group to perform. No coaxing, cajoling, or pleading is necessary. Everyone participates "because it is time" and because it is for the good of the whole tribe and the world.

The Zuni socio-religious system is composed of four separate but interrelated and interlocking systems; each operates independently and synchronically to provide for the physical and psychological

Figure 8.2. Morning prayer to rising sun.

---

*"He," "his," "him," and so forth are used throughout this chapter in the collective sense meaning "he/she," and so on.

needs of the users. Superimposed one upon the other are (1) the fourteen matrilineal, totemically named, exogamous clans (that is, groups claiming descent from a common ancestor and required to marry members of other clans, not their own); (2) the six men's groups, the kivas; (3) the ten curing or medicine societies; and (4) the two priesthoods, the Rain Priesthood and the Bow Priesthood. Around these four interlocking systems moves the annual religious and ceremonial cyclic calendar that holds the entire socio-religious system together.

Membership in a clan is not by choice, but by birth; however, non-Zuni members have occasionally been adopted by certain clans. This practice is not a common thing, but it is an accepted method of acquiring clan membership. The curing societies are responsible for healing various illnesses. Membership in them is based on accident, illness, trespass, or individual choice and is open to both men and women. Generally, membership in a curing society is for life. The Rain Priesthood is also open to men and women; however, membership is clan affiliated. Only men who have taken an enemy scalp may be members of the Bow Priesthood. Sometime between the ages of eight and twelve, all males undergo their final initiation into one of the six kivas. Women are not excluded from the kivas; however, because of the severe ordeal of initiation, they are not encouraged to join. In the 1930s, there was one woman member whose life had been saved by a kiva member, thereby obligating her to become a member of that kiva. In 1983, there were no women members in any kiva.

With the exception of the clans, each of the religious organizations is led by selected or appointed leaders. Leadership positions are not automatically inherited from uncle to nephew. This is true even for the clan-affiliated offices. For example, to hold the position of the Kiva or Dance Chief, an individual must be a member or a child of the Deer Clan; to hold the position of House Chief, one must be a member of the Dogwood Clan. Each officeholder is selected by the chiefly council on the basis of his character. Clan affiliation is important; it is the first criterion, but temperament is more important. Officeholders must be "good"; they must not be argumentative or stand out in public debate; they must not drink or be public nuisances; they must not fight; and above all, they must have "good hearts." These virtues make a "good Zuni."

The foregoing brief outline of organizational groups implies that a male tribal member has the option of membership in all groups and, at the very least, by birth he is a clan member. This membership in itself carries various kinds of religious obligations to his mother's clan and also to his father's clan. To the author's knowledge, no one has ever attained membership in all groups. However, almost everyone has peripheral knowledge of the functions of all groups and special knowledge of the functions of any groups in which he holds membership.

Women play a very special religious role. They must know the proper words of prayers to keep the corn maidens from running away. (In legend, the corn maidens ran away from Zuni because they were not respected and properly cared for. A great famine occurred until they were coaxed back to Zuni by the *nawi:que*, or galaxy society.) The women must also bless newborns with water and present them to the Sun Father with proper prayers. They must prepare the bodies of their deceased clan relatives for burial. They must prepare food offerings for the gods, feed the ancestors at every meal at the table or the fireplace, and also greet the morning sunrise. Men are responsible for the universe. Women are responsible for the family and the tribe. All life is traced through the mother. Thus, one's mother's household is the center of all major religious and ceremonial events.

Each unit of the Zuni socio-religious system functions independently and yet in concert with the others in a never-ending cyclic pattern. The coordination of the annual calendar was the responsibility of the Sun Priest, whose position has been vacant since before 1952, because of the exceedingly time-consuming and physically strenuous demands of the position. Now it is the House Chief of the Rain Priesthood who has this responsibility.

As in most agriculturally oriented cultures, the winter and summer solstice periods are the most ritually significant times in the Zuni religious calendar. During these two periods, everyone in the community participates in some way in numerous rituals.

Figure 8.3. Painting by an unknown Zuni schoolchild, c. 1900–1933, showing four *koyemshi* (mudheads).

The summer solstice marks the middle of the ceremonial calendar. At the end of June, the kiva group whose turn it is to make the pilgrimage to the sacred lake in the west makes preparations for the trip. The group is led by the Kiva Chief and the Fire-god and the Fire-god's ceremonial father. Also participating are members of two kiva groups who will perform the dances attended by the twelve men (two from each kiva) who are representing the Shalako gods for the year, and the mudheads—ceremonial clowns wearing earth-colored, globular cloth masks with knobs on them.

There are two sacred places that may be visited on the pilgrimage. One is at the junction of the Little Colorado and the Zuni rivers near the present town of Saint Johns in Arizona. This spot is the most sacred and is sometimes referred to as "Zuni Heaven." The other is a sacred artesian spring near the farming village of Ojo Caliente, New Mexico. After the pilgrimage ceremony is concluded, each kiva in turn performs a dance at Zuni as part of the summer rain dance series. This dance ceremony, which usually occurs in early July and marks the return of the dance gods to Zuni, is a prayer for rain. Synchronized with the summer rain dances is the retreat of each Rain Priest in turn to pray for rain.

There is much joy and delight when the rains come. The author saw such a downpour of rain during one of these summer rain dances/retreats that there was almost a river running through the central plaza of the pueblo. In spite of that deluge, the spectators never moved an inch and the dancers never missed a beat as they performed late into the evening.

The Rain Priests conclude their period of prayer and retreat at about the same time as the summer dance series comes to an end in early September. In October or early November, the preparations for the winter solstice are begun.

The winter solstice is the end of the old year and the beginning of the new one. All year long, the six kiva groups (who present the impersonations of the Shalako gods), the Long Horn (the head of the council of the gods), the council of the gods, and the mudheads have been building houses much larger than ordinary homes for those families who are their sponsors for the year. The construction is under the direction of the ceremonial representatives who have become ceremonial sons and family members of the sponsoring households. (After the sixth day of ceremonial activities, each of these "Shalako houses" will belong to its sponsoring family.)

The best-known event in the series of winter solstice religious activities is an all-night ceremony popularly called the "Shalako dance," or the "house blessing" ceremony. It is open to the public. Events leading up to the all-night dance start eight days before when the mudhead group goes into seclusion. Four days before the Shalakos arrive, the *saya:tash/a* (literally "Long Horn"), the council of the gods, and the kiva groups representing the Shalako go into retreat in their kivas. The men representing the various participants—the Shalako, the Long Horn, the council, and the mudheads—were appointed for this one year, so this is the period when their religious obligations for the year come to an end.

The night following the all-night ceremonies, the mudheads perform for a couple of hours in their ceremonial house, and then everyone rests.

Figure 8.4. Painting by an unknown Zuni schoolchild, c. 1900–1933, illustrating an initiation ceremony for young men.

The next night and the following four nights, there are dances by the six kiva groups in the Shalako houses. This is one of the periods when ceremonial dances are performed without masks. On the fourth night, there is dancing until dawn. On this night, those kiva members who for some reason did not perform in the regular dance series during the year have the opportunity to fulfill this obligation. At dawn, all kiva members who own their masks assemble in the ceremonial house and are prepared for their "departure" to the sacred lake.

The mudheads assemble in the central plaza and in the early morning are escorted to their clan's house. All the clan members with a clan brother representing a mudhead have collected "gifts," ceremonial offerings for the mudhead. Shortly before noon, the mudheads are escorted back to the plaza where all the offerings are collected. In the meantime, those members of the kivas who performed the dances in the plaza make four visits to the central plaza and exit eastward along the banks of the Zuni River. That night, the mudheads must visit each household in the village to do *a:lha/sh/ik/ya*, to "make old" or to bless with long life. These visits mark the end of this portion of the ceremonial cycle.

About ten days after these events, a period of time is set aside for the "fast,"* *itiwanan: te/chi/kya*, "arrival at the center place." Every man, woman, and child makes ceremonial offerings to the gods and abstains from certain foods for at least four days. This fast is done at the beginning of the new year. At this time, those who are not members of the priesthood or of a curing society and who hold no appointed religious office and therefore are considered "poor" make offerings of prayer meal and prayer sticks. Prayer meal, the basic offering on any occasion, is a coarsely ground, white corn meal. Sometimes bits of turquoise and seashell are combined with the meal.

Prayer sticks are made only by males. The young adult male who is a member of a kiva prepares two

---

*This is not a total fast, but certain foods are avoided. The nature of the fast depends on the individual's religious position. Everyone abstains from fats (meat) and salt for at least four days.

prayer sticks for the masked gods, two for the ancestors, and one for the Sun Father. The latter prayer stick is a willow shoot six to seven inches long to which are affixed the most precious feathers the maker owns. In the primary position on the stick, at the top, is a downy eagle feather, followed by a duck feather and as many colorful summer bird feathers as are available. The body of the stick is painted yellow. Very young, uninitiated males also make offerings, prepared by fathers or older brothers, to the ancestors and to the Sun Father.

Women, regardless of age, offer prayer meal and prayers sticks to the ancestors and to the moon. The prayer stick for the moon is the same as the one for the Sun Father except that it is painted blue. The prayer sticks offered by women are prepared for them by male relatives.

At the same time as the fast, the curing societies are active with their "free clinic." Once a year, the societies set aside one night, called "good night," so that anyone who wishes may be treated by the curing society.

The end of this period of the fast is sometimes referred to as the "new fire" ceremonies. During the fast period, people abstain not only from certain foods but also from trading, selling, buying, and visiting other homes, as well as from sex. In addition, no one carries or builds an open fire outside the home. All trash, thought of during this period as wealth, is collected in a shrine in one corner of the room marked off by a prayer meal enclosure.

During one of these days, clay is collected from along the banks of the eastern river. In the evening, individual members of the household make from it various objects, such as horses, deer, sheep, cows, cars, and money. All are modeled in miniature, as symbols of their makers' most fervent wishes for the coming new year. The miniatures are called /i/tsuma:we, or "wishes." The miniature objects are coated with ashes and placed in a crib in the "trash shrine" along with ears of corn.

On the morning of the ninth day, before dawn, the two gods of the Big Fire Curing Society bring the first fire into the village. In each family, the youngest male member has prepared a juniper bark fire brand, and as the gods pass on their way to the central kiva, he lights the fire brand from the family's fireplace. Then the father of the family bundles the trash from the shrine in a blanket, as if he is carrying corn from the field; the mother cradles the miniature objects; and the father and mother carry the ears of corn that were in the shrine. Led by the young fire carrier, they head for the nearby river bank. By this time, a hundred households are going down to the river, appearing like so many fireflies. The trash, the corn, and the miniature objects are neatly piled. All stand facing the east with a handful of prayer meal in their right hand. Each father gives a short prayer. At the conclusion, the prayer meal is offered by sprinkling it over the pile.

The ears of corn and the miniature objects are collected, and the families return home. The corn provides seed for the spring planting, and the miniature objects are "planted" later in the fields. (Some families leave the miniature objects with the trash, but the author's grandfather always brought his family's miniatures back and "planted" them later in the fields. All night long, the stew pot is going. When the family returns home, the fireplace is lit, a small portion of all the food is placed in bowls, and everyone stands facing the fireplace with the food. A short prayer is said, the food is put in the fire and the ancestors are fed, and everyone sits down and eats, and then the fast is over and the new year begins. About ten to twenty days later begin the winter night dance series and the plaza dances requested by the sponsoring families of the previous year. These dances continue until once again the visit to the sacred lake is announced.

The preceding are the highlights of the annual ceremonial events and religious life of the Zuni people, which are observed collectively. In addition to these group activities and paralleling the ceremonial and religious calendar are the individual actions that ensure for the person "a long road." Certainly, the Zuni believe that if they participate in these group events, they are being "good" Zuni and are adding to their life roads. However, Zuni religion—*tewusu*—is not limited to special times or places. It encompasses life every day, anywhere.

In the olden times and sometimes today, a person who observes tewusu rises before dawn and finds his way to the east of the village, where he stands by an anthill or a juniper or sage bush, facing the east with a handful of prayer meal. He greets the sunrise, asking for protection for himself, his family, and the people and asking that even though "some of his children" might wish to do harm to his people, this should not occur, so that they "may all reach the ends of [their] roads together in happiness." He then returns home to a prepared meal, places a portion from each dish in a bowl, faces the fireplace, says a short prayer, and puts the food into the fire to feed the ancestors and the gods. If there is no fireplace, a small pinch of food placed on the side of the plate is acceptable. This ritual is done at every meal. At the evening meal, a portion is placed in a bowl and taken later in the evening to the west edge of the village along the banks of the Zuni River, where it is offered to the dance gods and the ancestors.

The Zuni River is the spiritual lifeline of the Zuni people. Nearly every aspect of the religious system is in some way tied to the river. Along its banks and in the stream, offerings are made to the gods and to the ancestral spirits for continuous protection, spiritual guidance, and long life. In times past, the river was the absolute source that gave life; it provided drinking water for the people and the animals and water for the plants, which man and beast depended on for life in this high, dry plateau country.

The Zuni believe that the spirit returns upon death to the place where this muddy little stream converges with the Little Colorado River, northwest of Saint Johns, Arizona. The place is called

Figure 8.5. Carving depicting a Shalako *kokko* (Katsina), by an unknown artist from Zuni Pueblo before 1925.

*ko/lhu /wa/a la:wa*—"there to become one of the ancestors." The ancestors are said to return to *itiwann/a*—Zuni—by making their way back up to the Zuni River in the form of ducks whenever the masked dances are performed. They also return in the form of rain in the summer and snow in the winter to replenish the stream, to provide protection, spiritual guidance, and abundant harvests, and to give long life. The stream ties the living world—*ja/lona:/itiwann/a*—with the other world —*ko/lhu /wa/a la:wa*.

This brief account of the religious life of the Zuni is not a description in the past tense or an historical account. It is a narrative from the memory of one of the living members of the Zuni system. It is an account of a living, functioning, viable, vibrant culture that has withstood successive challenges and impacts with little change for more than seven hundred years. In the 1400s, the Athapascan-speaking bands came to Zuni land from the north; in the 1500s and 1600s, others came from the south from Spain and Mexico; and in the 1800s, settlers came from the east as the US westward movement brought "Western Civilization."

All of these had little or no effect on the religious life of the Zuni. Far greater effects on the people came from the diseases introduced by the migrants: among them, small pox, whooping cough, and measles. Subtle changes were made in the Zuni economic and subsistence systems as a result of the introduction of draft animals, wheat, melons, and fruits, as well as technologies such as carts, formed adobes, and metal tools. Along with these things came some adaptations in the Zuni language. But the basic political and religious foundation, to this day, has stayed the same. Even the underpinnings of the pueblo's present civil government are oriented around the Zuni system.

The late **Edmund J. Ladd** was an anthropologist, a University of New Mexico graduate, and a member of Zuni Pueblo. After a career as Pacific Archaeologist with the National Park Service, he joined the staff of the Museum of Indian Arts and Culture, Museum of New Mexico, in Santa Fe. He wrote this essay in 1983.

Figure 9.1. Motisinom (Basketmaker) pictographs in Atlatl Cave, Chaco Canyon.

# Yupköyvi
## The Hopi Story of Chaco Canyon
*Leigh J. Kuwanwisiwma*

*Aliksa'i.* Listen, let us begin. From the four cardinal directions they came. The Hisatsinom, the ancestral Hopi, were certain that a place called Yupköyvi, "the place beyond the horizon," was their destination.

The appropriate signs were there. The great blue star called the Sakwasohu—the supernova of 1054—had appeared in the heavens. This portent told the migrating clans to end their journeys and await further signs. Many clans had indeed fulfilled their covenant with Masaw, the spiritual guardian of the earth. This covenant dictated that clans journey to the four corners of the present world, place their footprints, and await further directions.

Masaw's guidance told the old ones that they had to begin a convergence on Yupköyvi, the place known today as Chaco Canyon (plate 15). It was a place where knowledge was to be shared and where the people would make final deliberations about their ultimate destination.

This is a part of their story.

### Antsa Yaw Yanhaqam Hiniwti: Yes, This Is the Way It Happened

It is well known among observers of the Hopi people that they continue to hold onto their traditions amid enormous strain. Indeed, it is amazing that Hopi ceremonies and traditions still play such a key role in Hopi life, enabling Hopi clans to understand and respect their individual cultural histories.

Archaeological evidence from Pueblo Bonito, for example, illustrates the use of ceremonial wands similar to those used today among Hopi religious societies. Scientific correlation of Hopi traditions with practices at places such as the ancient villages of Chaco Canyon does not surprise the Hopi people.

How did the people of Yupköyvi come to be? This is our knowledge about them.

The beginning of the fourth way of life was one of simplicity. The Motisinom, the "first people," whom archaeologists call Basketmakers, were planting, hunting, and harvesting wild plant foods in and around the valleys of the Chaco region. Song and social interaction were the spiritual forces that bonded these people. Their lives centered on their subsistence tasks and care of their environment. Among the Hopi clans descended from these people are the Katsina, Badger, Gray Badger, Tobacco, and Cottontail Rabbit clans. Today these people play important roles in the winter and summer katsina ceremonies.

According to oral traditions, these early ancestral Hopis occupied a vast territory in which they cultivated corn and squash, hunted game, and gathered edible plants. Their lifeway required cooperation and reciprocity. Clan groups lived side by side and traded with one another. At appropriate times, they played competitive games, among which long-distance running was popular. They were excellent storytellers and song composers. On occasion, the Motisinom held ceremonial dances that, although not elaborate or ritualistic, expressed contentment, happiness, and gratitude, especially in years of good harvest.

The first people of the Chacoan landscape were travelers and traders, too. They journeyed west to trade with the people of Koyongtupqa, or Canyon de Chelly, where the Eagle and Badger clans ruled supreme. To the south, they traveled to Tsipiya (Mount Taylor). Turning east, they visited their relatives along the Hopoqwvayu (Rio Grande), stopping to rest at the early villages of the Acomas and Laguna. To the southwest, the Motisinom made pilgrimages to the ancestral villages of the Zunis to trade for salt.

These early people were not free from environmental challenges. Drought was especially feared. Hopi recollections about the past speak of hardship and suffering. Hence, the people always practiced preparedness.

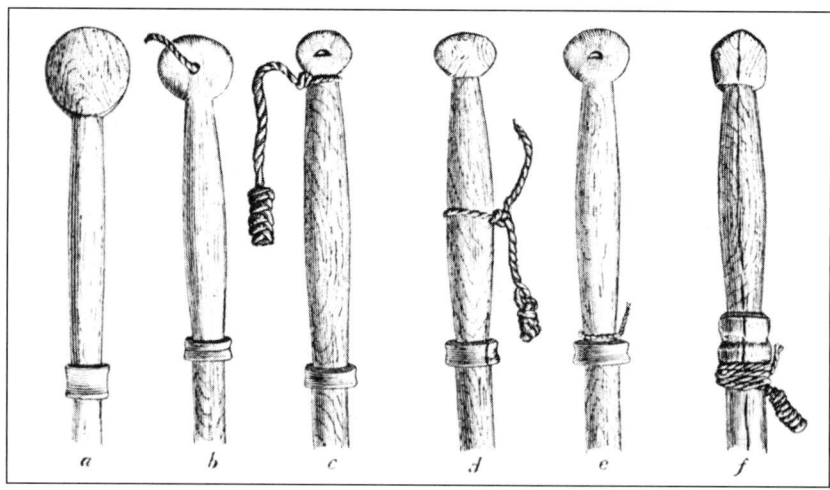

Figure 9.2. Ceremonial wands from Pueblo Bonito as illustrated in Pepper and Nelson's 1920 excavation report.

Hopis today have cultural teachings that reach back to these trying times. Visit any Hopi household and you will learn that it keeps stocks of traditional food crops such as corn that will enable its members to survive for a period of time. The stone granaries that are found in abundance throughout the Chacoan valleys attest to this necessary behavior, one still practiced by contemporary Hopi people.

In order to survive in a sometimes unpredictable environment, the Motisinom mastered adaptable technologies that assisted them in wisely using their sometimes limited natural resources. An understanding of weather patterns, for example, allowed farmers to devise planting strategies for the coming year. The Motisinom of Chaco also became astute observers of the cosmos. Their knowledge of astronomy enabled them to establish a calendar and ceremonial cycle and to keep track of mundane seasonal tasks.

All this, however, was early in the history of the Hisatsinom. This was the time of the Motisinom, the first people—the time before the great villages were built. Although their way of life seemed secure and the people content, events began to reshape the cultural setting the Motisinom had established.

Hopi clan traditions speak about other clan arrivals to the north, where many villages were being established. These clans had arrived from the west beyond the great canyon and from the southwest through Nuvatukyaovi, or the San Francisco Peaks. As the Hopi people say, "Yaw antsa naanani'vaqw öqiwta"—now from all directions they came. This emergence of more Hopi clans would forever change the cultural horizon of the first people.

But before we delve further into Hopi migration traditions, let us put them into the context of the modern-day Hopi perspective.

### *Itam Hapi Ngyamuy Amumi Toonawta*: Our Clan Traditions Are Our Roots

Western perspectives about the modern Hopis are riddled with so many opinions that it is easy to understand why historians, anthropologists, ethnographers, fiction writers, and, more recently, "New Age" believers continue to argue over who is right. Few researchers have bothered to ask the Hopis who they are. Those who did were treated with suspicion by Hopis and on some occasions deliberately given misleading information.

Only recently have researchers formalized their relationships with the very people they study. Formalizing means recognizing and accepting the Hopis' own research agendas and designs. Modern Hopi research is creative and does not arbitrarily dismiss science. Indeed, it seriously considers scientific findings and extracts information that corroborates Hopi traditional knowledge or is credible in terms of that knowledge. Although this may seem overselective, Hopis are not surprised that scientific conclusions complement their knowledge and verify

Figure 9.3. A Hopi bean field. The Hopi method of planting crops in sand dunes reflects methods used in Chaco Canyon in ancient times.

cultural continuity between themselves and cultures thousands of years old.

Outside researchers sometimes casually misuse the terms "Hopi" and "Hopi tribe" without understanding their meanings. In 1998, the Hopi Cultural Preservation Office of the Hopi Tribe took the National Park Service to task when officials at Chaco Culture National Historical Park determined that twenty-one tribes, including the non-Pueblo Navajos, were "culturally affiliated" with human remains excavated from archaeological sites in the park. The key issue was not just flawed research by park personnel, but the way they treated Pueblo people generically, as if they were a single cultural group. The Park Service failed to accept the uniqueness of each tribe and conveniently chose to lump them all as descendants of the early Chacoan people. This dispute was compounded by the park's acceptance of the Navajos' political claim of affiliation to this prehistoric group.

I make this point about the Park Service because Western researchers also have tended to treat the particular clans they study as generically representing all Hopi people. Instead, outside researchers need to look at Hopis within their cultural context in order to understand their history through Hopi eyes.

Today there are thirty-four living Hopi clans and at least thirty additional extinct clans. Fundamental, then, is to accept that there are sixty-four clan histories and, in many cases, specific clan religious beliefs and traditions. Ask any Hopi who he or she is, and you will receive the proud answer that the person belongs to a specific clan but is also a faithful practitioner of the Hopi philosophy of life.

This said, let us dive into the final chapter of Yupköyvi.

## *Itam Hapi Yaw Ang Kuktotani:* We Have Been Instructed to Place Our Footprints

Hopi people consider themselves stewards of the earth. However, they did not just proclaim themselves such. As clan histories relate, clans earned the privilege of being spiritual caretakers. This privilege originated in the migration era of Hopi clans, an era based on a spiritual covenant with Masaw, guardian of the fourth era of life. Clans from the north, west, south, and east began to move into the Southwest. The Motisinom of Yupköyvi had visitors.

The strikingly visible archaeology we see in the

Figure 9.4. Pictographs near Penasco Blanco in Chaco Canyon that may represent the supernova of 1054.

Chaco area and at other well-known places, such as Salapa (Mesa Verde), verifies the convergence of the migrating clans on these locations. The physical evidence reinforces Hopi traditional knowledge.

The covenant with Masaw charged the Hisatsinom to travel to the four cardinal directions, place their footprints, and await signs. Slowly, over perhaps two millennia or more, the Southwest became a mixture of different Hopi and other Pueblo clans and moieties that were embarking on deliberate journeys. Facing environmental hardships, the clans slowly placed their evidence to fulfill Masaw's spiritual pact. Today, the footprints of Hopi clans are defined as ruins, sacred springs, burials, landscapes, migration passages, artifacts, petroglyphs, and trading routes and trails.

To understand and express their history, Hopis rely on traditional knowledge and tangible practices such as song and ritual. For example, the appearance of the supernova of 1054—which, according to astronomers at the University of Arizona, was brightly visible for up to forty-five days—is today represented by the Blue Star Katsina, who routinely appears in the mixed Katsina dance. According to Hopi oral literature, this "blue star" was the supernatural sign to the Hisatsinom to end their migration and begin to converge on certain sites, including Yupköyvi (Chaco), Salapa (Mesa Verde), Hoo'ovi (Aztec Ruins), Kawestima (Navajo National Monument), Homolovi (the Winslow area), and Pasiwovi (Eldon Ruins, near Flagstaff).

Thus, Yupköyvi became a gathering place for

Figure 9.5. Ancestral Pueblo petroglyphs in Chaco Canyon.

clans from local areas and clans that stopped at what might be described as "staging areas" some distance away. Among the initial clans to settle in the Chaco landscape were the Parrot and Katsina clans. Later, the Eagle, Sparrowhawk, Tobacco, Cottontail, Rabbitbrush, and Bamboo clans arrived. Carefully, they were given places in which to establish their villages. According to tradition, this took time. Initial settlers became the ruling clans, which established order for the religious cycle and social responsibilities.

Together they contemplated their future. They shared their migration knowledge, spoke of the hardships they had endured, and cried out of sadness and

Yupköyvi 77

Figure 9.6. Interior of the great kiva at Aztec Ruins as reconstructed by the National Park Service.

joy. They learned to understand one another, even though they spoke different languages. Certain clans agreed that they must now prepare for the final journey to a place called Tuuwanasavi, the "earth center," which to the Hopis is their present home on the First, Second, and Third Mesas. Tuuwanasavi would be their final destination and their final home with Masaw.

But in order to be received and accepted by Masaw at Tuuwanasavi, they had to offer him something in return. This would be their respective bodies of clan and religious knowledge. This knowledge had to be complete and pure, for it was what each clan would ask Masaw to receive so that the clan could finally become Hopisinom—people of Hopi.

So each clan was allowed to establish itself at Yupköyvi. Each clan chose a matriarch and patriarch to lead it. The group collected its clan knowledge and incorporated it into its own ceremony. Then the clans announced that they would share their ceremonies publicly. Some clans were allowed to construct ceremonial kivas, where the elders guided their followers.

Finally, clan members performed their ceremonies. Some were elaborate, others simpler. All were performed and witnessed with the highest reverence. All the ceremonies were for the good of the people—for good harvests, rain, and the perpetuation of life. This went on for many years as more clans arrived.

As time went on, the Hisatsinom of the Chaco area made contact with people of other great villages and established new trading routes. On one occasion, the Bow Clan of Hoo'ovi (Aztec Ruins) was beckoned to perform its great Salako (Shalako) ceremony. The clan leaders consented. Anticipation rose, for the reputation of the ceremony was well known. Together with the Bamboo and Greasewood clans, the Bow Clan began its ceremonial preparations. This took years, they say. Finally, the trip to Yupköyvi happened. It was indeed a great sight as the Salako danced in the plaza. It rained.

The Bow Clan members stayed at Yupköyvi for a long time, performing their dance four times every sixteen years. Then they returned to Hoo'ovi, from where they carried the ceremony to Awatovi and later to Orayvi. The many kivas at Pueblo Bonito are seen by Hopis as Salako kivas because

Figure 9.7. Casa Rinconada.

every time the ceremony was performed, a kiva "home" had to be constructed for it (plate 8). Since the great Hopi Salako ceremony was fully resurrected at Hoo'ovi, the great kiva at Aztec is known to the Bow, Bamboo, and Greasewood clans as the Salak'ki, the "home of the Salako."

Another important ceremonial site in Chaco Canyon is the kiva known as Casa Rinconada. Its original purpose was as a site for the performance of the Hopi Lakon (Basket) ceremony, under the ceremonial sponsorship of the Parrot Clan.

Thus was life at Yupköyvi, until slowly the clans left for different places. The Hopi Mesas were one destination. Other clans went to Halona, today's Zuni. Still others went to Zia, Acoma, and Laguna. Some chose to stay for a while longer, until they, too, left.

Yupköyvi had served its purpose, and now it was proper to lay it to rest.

For the many descendants of the Hopi clans that once lived in and around Yupköyvi, it will forever be their mother village. It is perhaps a distant, mysterious place beyond the horizon, but it is a place that lives in their hearts.

*Pay yuk pölö.* This is the end of the story.

**Leigh J. Kuwanwisiwma**, director of the Hopi Cultural Preservation Office, is a member of the Hopi Tribe, the Third Mesa village of Bacavi, and the Greasewood Clan. He is a trustee of the Museum of Northern Arizona and serves on the Tribal Advisory Board of the Arizona State Museum. His personal interests include research into Hopi history, Hopi language preservation, and traditional Hopi farming.

Figure 10.1. Goats grazing in Canyon de Chelly below White House ruins.

# Canyon de Chelly

## A Navajo View

*An Interview of Mrs. Mae Thompson by Irene Silentman*

With the Navajo culture changing as it is today, much of Navajo life still revolves around storytelling. In the Navajo world view, everything that meets the eye reflects the presence of *diyin dine'é* (gods and supernatural spirits). The gods brought the Diné (the Navajo people) up to this world and formed Dinétah (Navajo land) for them. Many such stories are told in the wintertime by elders who sit around the fire with their children, grandchildren, and relatives. They explain how certain landscapes came to be as they are today. If there are other elders present, then they reminisce and compare and exchange stories. New versions are understood and fitted into the missing sections of their own stories, so the process is like putting together a very complex puzzle Many elders say that this is a way of cleansing and renewing your spirits.

To listen to these stories in the Navajo language is an experience of unmeasured beauty, and an English translation, even of high caliber, cannot do justice to the powerful narratives. Many times, the English translations do not capture all the special nuances of meaning expressed in the Navajo language. The following interview provides us with several of the many stories of significant events and incidents of Canyon de Chelly.

IS: *Tell me about yourself.*

MT: They call me Mae Thompson. I am from this place called Del Muerto. The canyon's Navajo name is Tséyi'. I do not know when I was born. An older brother of mine who died not too long ago used to tell me that we were born one day apart. He was my aunt's son. He used to tell me that we were close to a hundred years old, but my [social security] card says differently. It says I was born in 1908 in July. I do not know which is true. My mother gave me the Navajo name of K'ízízbaa'. This is the name I am known by in my community. My clan is

Figure 10.2. Mae Thompson in 1986.

Hanághááníí. I have two older brothers still living and five children, two daughters and three sons. There are also several grandchildren. My husband died several years ago. I live here at this house with my youngest son and his wife. My other children have their houses around close by.

*IS: Do you live here all year round?*

MT: No. We live here only part of the year. Our summers are spent in the canyon. We have houses in the canyon that is called Black Canyon. The place we live at is actually near the mouth of Canyon Del Muerto and Black Canyon at a place called Tséláán. Our winters are spent up here.

*IS: When do you move down into the canyon?*

MT: It depends. Presently, our move into the canyon is determined by the water flow in the canyon. We used to spend only the winter season up here and move down into the canyon by spring, when we only had horses and donkeys as our means of transportation. Now our move depends upon the water flow. You see, they open up Tsaile Lake and Tsééhats'ózí Dam in the spring, and the water continues to run until midsummer. This is the reason why we do not plant until midsummer, which is really late. When this happens, then we do not harvest as much corn anymore, and we do not move into our homes until midsummer, when the roads are dry enough to travel on. We enter the canyon now through Chinle. It is not like the old days anymore when we could use horses and donkeys. Now, we have trucks, so we have to drive around Chinle. That is the way we also haul out our harvest.

*IS: What do you plant in the canyon?*

MT: We plant corn, watermelons, pumpkins and squash, beans, and hay. We also have an orchard of peaches, apples, apricots, and plums, in addition to several grape vines. When it rains, then we have an abundant harvest. Otherwise, it is usually small.

*IS: How long have you and your family been living in the canyon?*

MT: Ever since I can remember. We did not always live at this place where my house is now [*on top*]. We used to live closer to the edge of the canyon near that hill [*in the southeasterly direction from her house*]. One can still see the remains of our homes there. I can recall when I was a child. We were living at the bottom of the slope of that hill, and it was around harvest time when I saw several packed donkeys come up the horse trail with that fall's harvest. They were so loaded down with sacks and sacks of broken pumpkins. People used to do this at that time. They would break up all the pumpkins, pack them in large sacks, then load them onto donkeys to bring them on out to the top. Doing this made it easier to bring out the harvest.

*IS: When did you move to this area here?*

MT: My mother and younger sister were still alive when we moved here. I had just begun having children at that time. That was many years ago.

*IS: Did your parents ever tell you any stories about the canyon, events that occurred in the canyon or stories of people who used to live there long ago?*

MT: Do you mean the burnt homes in the canyon?

*IS: Yes, that plus stories of how people used to live long ago.*

MT: Those homes in the canyon were not burned by the Diné. Those ruins are Anaasází [*Anasazi*] homes (plate 21). It is said that when the Anaasází lived there, a big tornado came and destroyed them. This tornado came into the canyon from Chinle. It was a big whirling wind with fire. It went up each canyon and burned all the people. One can see these burnt areas today. They are those black bands and streaks on the cliff walls. They became this way from the fire and smoke. Everything in the canyon was destroyed, even the animals and all the

Figure 10.3. Canyon de Chelly, showing Spider Rock.

vegetation. It is said that this whole area and the canyon were very thick with all types of vegetation and trees. The vegetation was woven with the trees, which made it very dense. It was lush and wild and jungly.

*IS: Why were the people destroyed?*

MT: They began to do and learn things beyond the knowledge that was set for them. It's like what is happening today. People began doing many abstract things, drawing and painting. Things became so abstract and intangible. This is why they were destroyed. Some of their paintings and drawings are still on the cliff walls. They made drawings of the wind, air, and all kinds of animals like cows, deer, buffalo, elk, and birds. They obtained knowledge beyond what was set for them. These are the stories of my father and my grandfather. My grandfather's name was So'nii'biye' [*Son of the Star*].

When I was about five years old, I remember, Bilagáanas [*white men*] came into the canyon and began excavating, going through the Anaasází ruins. They dug up all sorts of things like clothing and bones of people and animals. I used to see all the things these Bilagáanas would dig up. They would lay them down for inventory before they would haul them out. It's amazing how the Anaasází lived. Their clothing was made of buckskin sewn with sinew, and they wore a lot of turquoise, chunky necklaces. I also saw combs [*bé'ézhóó'*], grinding stones, turkey skeletons, and pottery of various sizes. I don't know where the Bilagáanas took all their findings, probably back to their homeland. Yes,

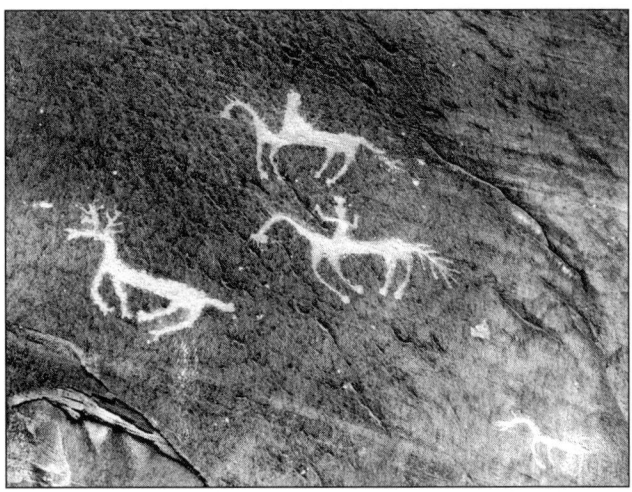

Figure 10.4. Petroglyphs of hunters chasing a deer, Canyon de Chelly.

Figure 10.5. Detail of Navajo rock painting depicting Spanish horsemen, Canyon del Muerto.

the Anaasází were destroyed because of their knowledge. They were afraid and built homes in the cliffs and walls of the canyon, but the gods did the worst that could be done to anyone—they destroyed the Anaasází with fire.

*IS: You say you live at a place called Tséláán in the canyon. Are there any stories related to this area that your kinfolk have told you that you want to share with us?*

MT: Yes. After the destruction of the Anaasází, the Diné moved in. They were living there happily until the Bilagáanas came in. The Bilagáanas would station themselves along the rims of the ledges and shoot at the Navajos. Many Navajos were killed, and their bodies would just lie there on the canyon floors. This is why they took refuge on Tséláán. My grandmother, Asdzáán Das, was one of the ones who took refuge up there. There was another woman, called Asdzáán Tséláán, and many men who took refuge there, too. When the People climbed to the top of Tséláán, they dug an underground pit for hiding. When you see the top of Tséláán, there is no place where you can dig because it is all a layer of rocks, except for this one place. At the top, there is a gap. In this gap is a small hill, and this is where the People made shelter and hid from the Bilagáanas.

Shots were coming from the different bluffs and ridges, and the People did not know what to do. A man named Dibé Yázhí Bich'ahí [*The One with a Lamb Hat, so named because he wore a hat made of the skin of a newborn lamb*] had an idea to trick the Bilagáanas. He put his hat on a long stick and waved it around from different positions, and the Bilagáanas would shoot at the hat. He did this so the Bilagáanas would use up their bullets. When this trick did not work too well, then Dibé Yázhí Bich'ahí and another man, named Ch'il Haazhiní [*Black Weeds Extend Out*], decided to use witchcraft on them. They said if they did not do something, then all the People would be killed.

Across the canyon were some box elders. The Bilagáanas camped here. It seemed as if they were going to be there a long time, because they built a stone structure under the box elders. They were going to subdue the Navajos by starving them. So it was that the two men began to perform the Witchcraft Way ceremony at dusk. They continued the ceremony throughout the night. During the night, one of the Bilagáanas yelled out, then another one, and

Figure 10.6. Aerial view of Canyon del Muerto (*left*) and Canyon de Chelly.

another. Pretty soon, all of them were making noises, yelling, whooping, and cursing. It turned out that the captain was killed [*witched*] first. The following morning at dawn, a young Navajo was sent to see what was happening at the white men's camp. He saw the men packing their horses, mules, and donkeys. They were leaving, so he watched them until they left the canyon. The Navajo ran back to the others and told them the happy news.

IS: *Who were these Bilagáanas? Were they cavalrymen?*

MT: Yes, they were cavalrymen. They wanted to kill off all the Navajos so they could take this beautiful land for themselves. It is said that they wanted to live here. However, they did not kill us all. Many Navajos hid in the high cliffs.

IS: *How did the Navajos get to these high cliffs? It seems almost impossible to climb up the sheer walls.*

MT: They used ropes to get down to the small cliff alcoves. They would put a stake at the edge of the cliff and use sash belts or yucca ropes to lower themselves to these alcoves. As they lowered themselves down to these alcoves, they made hand and toe holes [*steps*] with a rock called *nił*. This rock is a hard rock and was used a lot for hammering or digging. The hand-and-toe trails were made so the people could climb back up. Of course, they did not climb back up until all the Bilagáanas were gone.

IS: *What kind of food did the Navajos have with them in their places?*

MT: Not much. The one food that kept the Navajos from starving is called *ts'áálbáí* [*shelled, steamed corn dried and ground to powder*]. They would have just a handful of this, compacted in lambskin pouches. The women carried this food with them all the time. During the times when the men went hunting and brought back meat, they would dry the meat and grind it to powder, too. This is what the men carried with them. They had very little food.

IS: *How did they get water up in those cliffs?*

MT: They used large gourds for water bottles. The tip at the tail end would be cut open, and the gourds would be scraped out that way. When the gourds were all cleaned, then they would cut pieces of buckskin to cover the openings. This is how they kept water. There is a story that after spending several months living in the cliffs, the People ran out of water. There was no water to be found on top of Tséláán, so they had to go down into the canyon to get water. They used the ropes of sash belts and yucca to lower themselves down into the canyon at night. They filled all their water bottles with water at the water hole in the canyon and brought them back up. This is how the People overcame thirst.

*IS: Did this event take place when the white men came and burned all the peach orchards and corn fields?*

MT: Yes, that's right. This all took place around that same time. The white men burned all the orchards and fields. Many Navajos hid in the mountains—the Chuska Mountains. The white men rode up after them, but they were killed. The Navajos used arrows to kill them. They would climb on the trees and ambush the white men who rode past them. Many white men were killed with just bows and arrows.

*IS: Is this the time the Diné were marched to Hwééldi [Fort Sumner]?*

MT: Many events, including the stories I just told you, took place before Hwééldi. There is another story that occurred several years before. The Naakaii [Mexicans] used to come and steal women and children and/or capture Navajos for slavery. Those captured would be taken back to Mexico. One woman returned from there.

She was called Naakaii Asdzáán [Mexican Woman]. Her mother was captured as a child, taken to Mexico, and was impregnated by a Mexican. Naakaii Asdzáán was born. She grew up to be a beautiful woman, light pinkish complexion and curly hair. She is now dead.

Soon after her mother died in Mexico, Naakaii Asdzáán became very homesick for her mother's homeland, so she escaped and started back. As a young woman, she was also impregnated by a Mexican boy, so there was a baby girl. She started off, carrying her child on her back, piggyback-style. After ten days of walking, she saw that she was not gaining much ground and was becoming very weary and tired. She did not have any food with her except a small bag of ts'áałbáí. She drank this with water, you know, just enough to keep up her strength and keep her going. After ten days, she could no longer carry her child anymore, so she threw the baby on the ground and stepped on her throat to choke her. The baby was crying at first. Then she stopped. Naakaii Asdzáán took her foot off the baby's throat and started off on her journey again. When she was a few feet away, she heard the baby let out a faint cry again. She returned to the baby, saw that it was still alive, and she just could not see her baby die there. She picked up the barely alive baby, fixed some ts'áałbáí, and fed it to her. Then they began their journey again. That baby grew up to be a beautiful woman just like her mother was. She also is no longer alive, died an elderly woman. They used to live down there towards Chinle.

When the Diné were marched to Hwééldi, they were being blamed for raiding and stealing. On this long walk to Fort Sumner, some people walked, some rode on wagons, and some rode horses. It took many, many days before they arrived there. People spent several years there. Many wanted to return to their homeland, but the Bilagáanas would not let them. People begged and cried to be released to return home. The Bilagáanas are a mean people. Many Navajo women and young girls were impregnated by the white men, both Naakaii and Bilagáanas. When the People were finally released and started back, they had a lot of half-breeds with them.

*IS: Did your parents go on this long walk to Fort Sumner?*

MT: No, my father did not go there. These are stories that were told to him. My grandmother was one of the ones who went there. There are some trees in the canyon where we live that just fell last summer. It is said that the People were taken to Fort Sumner when those trees were very young. A couple of them were still green and standing when we left last summer. Maybe they will fall this summer.

*IS: Are there any other stories that you would like to share with us? Perhaps you know of some stories of how the canyon [Canyon de Chelly] was formed.*

MT: Yes. You did not ask me that, but I will tell you. My maternal grandfather said this story was told to him. A long time ago, the canyon was not there. There are two versions of this. Some say the canyon was not there, others say it was already there. I will tell you what my grandfather told me. There was a big lake that formed by itself at Tsaile. In this lake lived a big monster called Deelgééd [*a mythological monster similar to a rhinoceros*]. I do not know what it was, but people have seen it. It would roar in a thunder-like manner, saying "Diil-l-l, diil-l-l." People say it did this when it was mad. One day, it became very mad, probably at the People. The monster plowed open the lake, and the water rushed out and formed the canyon. This appears to be a very reasonable explanation for the formation of Canyon de Chelly, because one can see the water level that formed on the canyon walls. Today, it is still like that. You can see where the water level was. I think it's true. When the monster plowed the lake open, the water made its way down toward that west mountain. Another lake formed there at the base of that mountain. The lake is called Tódiníhí [*Water That Roars*]. The monster's roar was heard there again. It lived there for several years. Then it moved on to Bilagáanatah [*white man's land*]. I think this is true, too, that the water rushed down into Chinle and to the base of that mountain, because one can find those smooth water pebbles all along towards Chinle and all around there. There is a path that you can follow that leads to Tódiníhí Lake, and you find these pebbles all along there. It is said that the monster moved on to Na'ní'á Hótsaa [*Navajo Bridge, Arizona*]. Now, people hear it roaring there, and we do not know if it will plow open that dam. However, no one has actually seen that monster in that area. I don't know what will happen.

The home of Mrs. Mae Thompson and her family at Del Muerto, Arizona, is located across from Highway 64 above Antelope House Ruin and Standing Cow Ruin. Her children live nearby. She raises a few sheep and goats and has a small peach and apricot orchard. She also plants a small garden in the spring and grows several varieties of wild flowering plants. She planted the peach and apricot trees herself. The oldest tree is about three years old, and she says that the fruit on this tree is very sweet and juicy. She informs me that many people buy or trade for the trees she plants. It makes her feel good that people come to her for her trees, because she feels she is contributing to their economy and welfare.

With her stories, Mrs. Thompson has provided us with another perspective—that of her people, the Diné. She has enlightened us with stories that have been passed down from generation to generation. Her stories illustrate to us that the world we live in was shaped by great supernatural forces and that if we do not maintain harmony and balance with the supernaturals, the gods will become angry and alter the landscape again. In the stories the Navajos tell, there is always a moral to be learned.

**Irene Silentman**, a former assistant professor of languages at Northern Arizona University and specialist in bilingual education, is a member of the Navajo Nation. She is a translator and teaches Navajo linguistics through the Navajo Language Academy. Her clans are Hooghanlani and born for Tl'ashchi'i with maternal grandfather as To'aheedliinii and paternal grandfather as Naakaii dine'e. She lives in Newcomb, New Mexico, where she helps run her father's farm.

**Mae Thompson** is an elder member of the Navajo tribe and a resident of Canyon del Muerto, Arizona.

Figure 11.1. Peshlakai Etsidi, 1936.

# The Wupatki Navajos

## An Historical Sketch

*Alexa Roberts*

Tuba City's damp dark woolen mill was an old sandstone building that served as a temporary Protestant mission. There, in 1897, the Reverend William R. Johnston first met Peshlakai Etsidi, whose Navajo name meant Silversmith. Struck by the Navajo's outstanding appearance and dignified manner, the reverend inquired about him and learned that he was an apprentice medicine man and one of the first Navajos to learn silver jewelry making. In his community, he was admired and thought to have potential as a wise leader.

When Wupatki National Monument was established by presidential proclamation in 1924, the lines that demarcated its boundaries just south of Black Point were drawn over the territory held by Peshlakai Etsidi and his children and grandchildren. Those grandchildren, great-grandchildren, and great-great-grandchildren inhabit the Wupatki Basin to this day, a century and a half after Peshlakai Etsidi first settled it. The Navajo story of Wupatki begins with him.

Peshlakai Etsidi's predecessors first came to the Coconino area between Wupatki and the Grand Canyon during the late eighteenth century, sharing the resources offered by the Coconino Plateau and the Little Colorado River with Havasupai people throughout the first half of the nineteenth century. These early Navajo families were nomadic, traveling over immense areas each year to gain a subsistence based on sheep herding, agriculture, small game hunting, and wild plant gathering. Except for occasional skirmishes with Mexicans, Utes, and Hopis, the Navajos' seasonal travels were uninterrupted until the arrival of the US Army in the early 1860s.

Under the orders of US General James Carleton, in 1863 Colonel Christopher "Kit" Carson launched a military campaign against the Navajos, the ultimate aim of which was to confine the entire Navajo population at Fort Sumner, New Mexico. Males resisting US troops were to be killed, and women and children taken captive. A bounty was set on

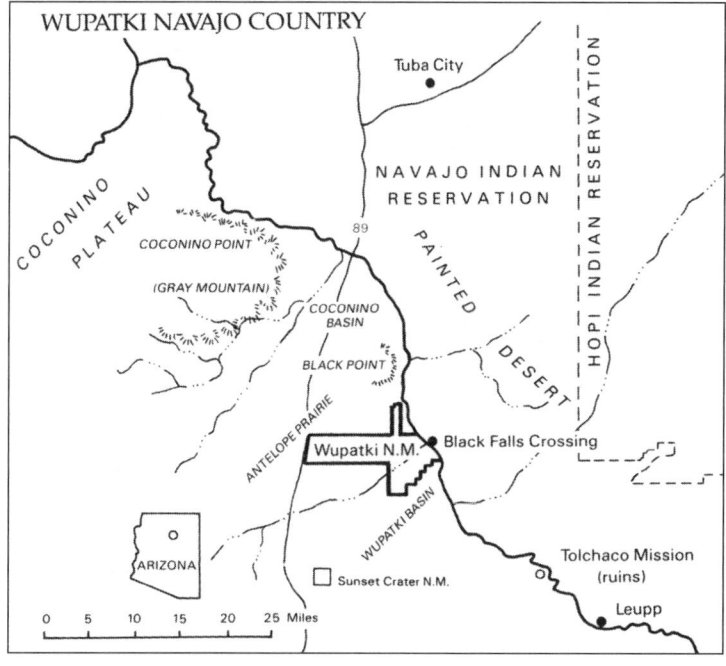

Navajo livestock, and their crops were to be destroyed. Word of the campaign reached the Coconino area in 1864, and the majority of Navajo families took flight, seeking refuge in the Grand Canyon and other inaccessible locations. The canyon settings of archaeological sites near Wupatki dating to this time suggest the Navajos' defensive position during this traumatic period.

By 1867, faced with starvation, local Navajo leaders convinced the Coconino Navajos to surrender to US troops. Peshlakai Etsidi, then about eleven or twelve years old, along with his family and other families, including that of his future wife, began a five-hundred-mile journey from the Grand Canyon to central New Mexico. The aged and infirm died along the way, and many more died during the years of captivity at Fort Sumner. Not until the utter failure of the "settlement" program in 1868 did Navajo leaders and the government sign a treaty that established the boundaries of a small reservation and allowed the Navajos to return to their homes.

Peshlakai Etsidi returned with his family to the Coconino Plateau in about 1870. Shortly thereafter, he married and moved to his wife's family home at Black Point, three miles north of Wupatki. During the next two decades, Peshlakai Etsidi and his wife established a family, acquired large herds of livestock, and worked a ten-acre field of corn. Other Navajos also returned to their original homes along the Little Colorado River and rebuilt their decimated herds of horses and sheep and their agricultural fields. Navajo populations spread out east and west of the Little Colorado River, from south of present-day Leupp to the Grand Canyon, including what is now Wupatki.

Before the century was over, Anglo incursions into the far western Navajo territory began again. The year 1876 brought Mormon settlers to the Little Colorado; they established colonies along the length of the river and crossed frequently at Black Falls on Wupatki's eastern border. In 1882, the railroad reached the newly established town of Flagstaff and began bringing pioneering Anglo settlers who would shape the history of the West. The US government granted the railroad company alternate one-mile-by-one-mile sections for forty miles on

Figure 11.3. The newsletter of the Tolchaco Mission, 1903.

either side of the track right-of-way. These railroad sections could be sold by the railroad company or traded for public domain land elsewhere.

Newly arrived Anglo ranchers coveted the scarce water sources and abundant grass on the railroad sections and public domain lands, some of which were occupied by Navajos. By 1897, conflicts had developed. In a typical incident, the board of supervisors of Coconino County, which included local Anglo ranchers, ordered a twenty-man posse to "assess the property" of Navajos in Coconino County. The board demanded a five-dollar tax per hundred head of sheep to be paid immediately by sixteen Navajo families on the Coconino Plateau. The supervisors knew that they would be unable to pay, and when the money was not received, the posse drove both the people and their sheep from their traditional grazing land. In a September 10, 1897, report of the Office of Indian Affairs, the

Figure 11.4. At Tolchaco Mission, 1904. *Back row, left to right:* F. G. Mitchell; Vera Standish, Peshlakai Etsidi's wife; Peshlaki Etsidi; Reverend William R. Johnston; Bwoo Adin, one of the headmen who went to Washington. *Middle row:* Lizzie Scott, the mission teacher; Mrs. W. R. Johnston; Philip Johnston; David Johnston. *Front row:* Mary Johnston, La Pah.

commissioner of Indian Affairs noted:

> Snow was falling (a deep snow already covered the ground), the weather was bitter cold, and the ewes were lambing. The Indians pleaded for a reasonable time within which to remove, but were denied. Their houses and corrals were burned and they and their flocks were rounded up and pushed toward the Little Colorado River with a relentless haste, the posse keeping women, children and animals in a fright by an intermittent fire from rifles and revolvers. When the river was reached it was found to be so deep to require the sheep to swim. The posse surrounded the flocks and pushed them into the water, and nearly all the lambs, with many grown sheep, went down the stream or chilled to death after crossing, and many died afterward from the effects of exposure. The loss to the Indians was equivalent to several thousand dollars.

Peshlakai Etsidi's family was wintering on the Coconino Plateau that January of 1897. Although offenses against their people had happened before, Peshlakai Etsidi and other influential leaders urged the Navajos not to seek retaliation against the Anglo strangers. Instead, on behalf of the Coconino families, Peshlakai Etsidi traveled to Tuba City to request the assistance of the newly arrived Protestant missionary, Reverend William R. Johnston. Their meeting was the beginning not only of a lifelong friendship but also of a small influence on Indian/government relations. Peshlakai Etsidi spoke to Johnston with a power and eloquence that moved Johnston to action on behalf of the Navajo people. "In-a-way," Johnston remarked in retrospect, "it was like meeting one of the biblical figures, Isaac or Jacob."

Three years later, Reverend Johnston, his wife, and their children, Mary and Philip, moved from Tuba City and established a mission at Tolchaco, about fifteen miles upriver from Wupatki's eastern boundary. In 1902 and 1904, Peshlakai Etsidi

Figure 11.5. Katherine Bartlett (*far left*), Dr. Harold S. Colton (*seated, right*), and Dr. and Mrs. Will Dakin (*right*) visiting the Peshlakai Etsidi family in Wupatki Basin, 1936.

accompanied Johnston and other local headmen on two trips to Washington, DC. There, with young Philip Johnston acting as interpreter, Peshlakai Etsidi presented the story of Navajo hardship to President Theodore Roosevelt. The president listened, and after presenting Peshlakai Etsidi and his companions with medals proclaiming them official leaders of their people, he instituted an immediate program of reservation boundary extensions and Indian allotments in the western territory. Government land surveyors and allotting agents congregated in the Coconino area during 1908 to conduct land studies for the distribution of allotments and additions to the reservation created by the 1868 treaty. The twenty-four-square-mile Leupp extension, about twenty miles south of Wupatki on the Little Colorado River, was officially added to the reservation. This extension enabled the addition of a school and irrigation projects, as well as the protection of some Navajo grazing range from Anglo ranchers.

Thus, within the first decade of the twentieth century, presidential attention was drawn to the plight of the Navajos on the public domain in the Wupatki area. But presidential attention turned to the area for other reasons as well. Severe looting of major ruins by local pothunters prompted Dr. Harold S. Colton, founder of the Museum of Northern Arizona, to request government protection of the two largest ruins. Three and a half sections of land were set aside as Wupatki National Monument under presidential proclamation 1721 on December 9, 1924. Ten years later, the monument received its first full-time custodians, Jimmy and Sally Brewer, who lived in two converted rooms of the ancient Wupatki Pueblo. To cover the forty miles to Flagstaff on dirt track, they used a truck that the National Park Service purchased from the Bureau of Public Roads for five dollars. The Brewers quickly established close friendships with their Navajo neighbors.

The 1980s were rough for the Navajos, as they were for the entire nation. Tuberculosis was rampant, and the local families sacrificed much of their property to pay healers who performed curing ceremonies. After each death, the hogan in which the person died was abandoned, according to custom,

Figure 11.6. Clyde Peshlakai, acting custodian at Wupatki National Monument, 1936.

leaving some families temporarily homeless. The fervor of the 1908 allotting program had subsided, and many government promises remained unfulfilled. The grazing areas that were left to the Navajos by ranchers were poor and severely overgrazed, and the remaining sheep were unprolific. Yet, despite the lean economic situation, the friendship and mutual respect established between the Brewers and the Navajos brought Anglo/Navajo relations at Wupatki to an all-time high.

The pinnacle of these good relations between the National Park Service and the Navajos occurred between the winters of 1935 and 1936, with two legendary Christmas parties and the first Navajo Craftsman Exhibition, now a renowned annual function of the Museum of Northern Arizona. These events involved the Brewers, representing the National Park Service; Dr. and Mrs. Harold S. Colton; Harvard archaeologist Watson Smith; archaeologist Charles Steen; historian and archaeologist Katherine Bartlett; and Philip Johnston and his wife, Bernice; in addition to about forty Navajo residents of the Wupatki Basin, nearly all of whom were Peshlakai Etsidi's children, grandchildren, and in-laws.

These events not only made the Navajo occupants of the Wupatki Basin an integral part of the National Park Service operation but also brought regional attention to this remote corner of Arizona. The Navajo Craftsman Exhibition was held in a specially constructed ceremonial hogan and four large ramadas just north of Wupatki Pueblo. The event required four months of joint preparation by the Brewers, the Coltons, and all of the Navajo residents of the Wupatki Basin. On June 6 and 7, 1936, there were 127 visitors who traveled the forty miles of dirt road from Flagstaff to Wupatki to participate in the significant event.

The show perpetuated knowledge of vanishing traditional arts among the Western Navajos, the effects of which are carried on in the work of today's Navajo artists. It set the stage for the Museum of Northern Arizona's annual Navajo Craftsman Exhibition, which draws Navajo artists and visitors from around the country and has provided a starting point for many Navajo artists. Most important, however, the Navajo show and the two Christmas parties at Wupatki in 1935 and 1936 created the first really successful relationships between the Wupatki Navajos and Anglos since their first contact almost eighty years' before. At these functions, Peshlakai Etsidi gave his last speeches, urging continued cooperation between the two cultures that had become permanent residents of the same high desert.

Peshlakai Etsidi was buried in the hogan in which he died in April 1939. Before his death, Wupatki National Monument was enlarged from 2,234 acres to nearly 35,000 acres, encompassing the entire Wupatki Basin. Dr. Colton and Philip and Bernice Johnston continued to work for the Navajos' rights to their traditional homelands, emphasizing the role of the National Park Service in protecting the Navajo culture within Wupatki's boundaries. The Navajos also worked towards the cooperative relationships that Peshlakai Etsidi had urged. During the 1930s and 1940s, for example, Peshlakai Etsidi's son, Clyde Peshlakai, played a

Figure 11.7. Descendants of Peshlakai Etsidi. *Left to right:* Della Yazzie, a granddaughter; Helen Davis, a great-granddaughter; Stella Smith, Helen's mother and Della's sister; and three great-great-grandsons, Aaron, Tony, and Myron Davis.

fundamental part in Park Service operations. Clyde became Wupatki National Monument's "support staff." He conducted tours; participated in the construction and design of facilities; maintained the grounds; provided the custodians and visitors with folk knowledge of the monument's natural and cultural features; and in 1956 even located á unique geological feature, the blow hole, near the Wupatki ballcourt.

By the second half of the 1940s, however, Navajo and Anglo attention turned from local concerns to World War II, in which the Wupatki Navajos indirectly played a significant role. Philip Johnston had learned fluent Navajo while growing up at Tolchaco Mission, and he organized the Navajo Code Talkers in 1942, forty years after interpreting the conversations between President Roosevelt and Peshlakai Etsidi. Using a code language based on Navajo, the Code Talkers played a significant role in US military communications during World War II. "Considering the remarkable fact," wrote Johnston during the war, "that we have the only system of air-tight, secret communication by voice-radio, I have every reason to believe that the Navajos will make real history before we are finished." Johnston's prediction proved correct.

The Wupatki Navajos fought in the war as well, and many young men enlisted in the army to serve the nation, less than eighty years after the US Army had imprisoned their grandparents at Fort Sumner. The war was far from Wupatki, but it hit close to home as Navajo mothers and fathers asked the superintendent of the monument to write letters to their sons stationed in foreign places of which they had never heard.

Increasing governmental bureaucracy following the war years eventually strained relations between the Navajos and the National Park Service. By the early 1960s, all Navajo families except that of Clyde Peshlakai had moved off the monument. Clyde died in 1970 and is buried in the rock house just north of monument headquarters, off the monument highway. Today, his offspring—third, fourth, and fifth generation descendants of the original Navajo settlers—continue nearly one hundred fifty years of Navajo heritage within Wupatki's boundaries. Wupatki National Monument protects the numerous archaeological remains of this long Navajo occupation. The archaeological record represents five generations of a Navajo family whose experience at Wupatki reflects not only the course of local events but also broader manifestations of the American Indian role in the nation's history.

**Alexa Roberts** is superintendent of Bent's Old Fort National Historical Site and Sand Creek Massacre National Historical Site. She is a former staff member of the Navajo Archaeology Department. She wrote this essay in 1987.

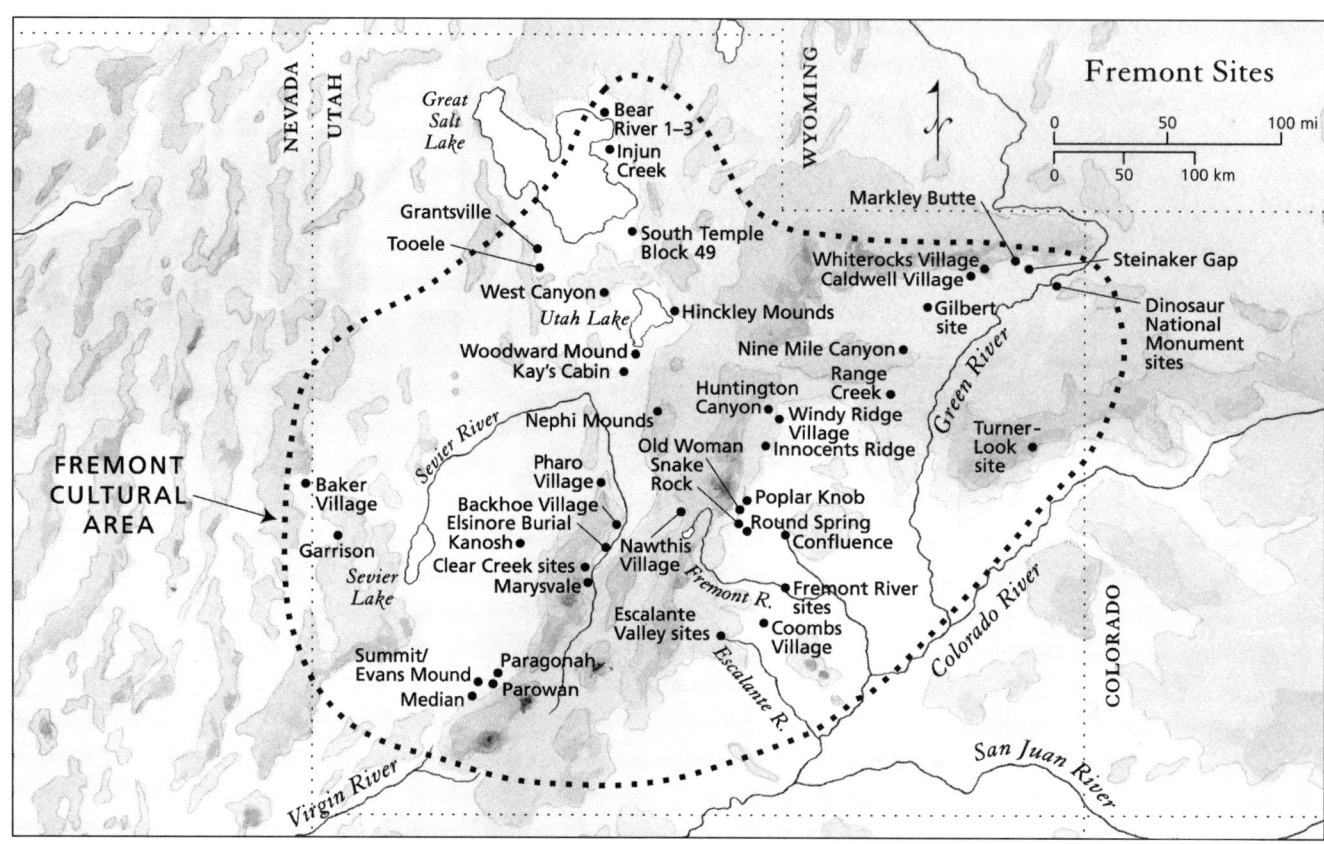

Figure 12.1. The Fremont culture area at its greatest extent. Horizontal scale is exaggerated.

# The Enigmatic Fremont

*Joel C. Janetski*

Noel Morss, a budding archaeologist on a 1929–31 Harvard-sponsored expedition into Utah, explored caves and alcoves in the picturesque canyons of the Fremont River and its tributaries in south-central Utah. He dubbed the "partly or predominantly agricultural" people whose remains he excavated the Fremont, after the river. Morss recognized that these people showed many similarities to ancestral Pueblo Indians. The Fremont people grew corn, beans, and squash, made pottery, and constructed adobe and masonry storage features, just like the ancestral Pueblos. Yet, they were different from those more southerly peoples—they wore moccasins, not sandals, made basketry in a different way, crafted large numbers of clay figurines, and seemed to rely more on wild foods.

Although Morss first used the term *Fremont* and recognized that this distinctive culture was widespread, he followed in the footsteps of several pioneers of research into these people. Edward Palmer had explored what he called Pueblo sites in Utah in the late 1800s and sent collections from them to eastern museums. Henry Montgomery, a professor at the University of Utah, had brought a more systematic approach to investigating Fremont mounds in the 1890s. Morss also acknowledged earlier work by Neil Judd in the 1910s at mounds in central Utah along the westward-facing Wasatch Front.

Julian H. Steward followed Judd and Morss in exploring Fremont culture while teaching at the University of Utah in the early 1930s. Like Judd, Steward mostly dug in mounds along the Wasatch Front and labeled the sites Puebloan, focusing on similarities in ceramics, architecture, and corn farming with ancestral Pueblo sites to the south.

Morss's term and the idea that these sites represented a distinct culture won the day, thanks to the research of Jesse Jennings and his students. Jennings arrived at the University of Utah in 1947 and started an aggressive program of archaeological research that included caves and rockshelters, as well as mounds. Working closely with graduate students in the 1960s and 1970s, he defined the geographical extent of Fremont culture and, through radiocarbon dating, placed it in time. This work led to abandoning "Puebloan" in favor of Morss's "Fremont," in recognition of the latter culture's uniqueness.

Fremont, then, refers to a cultural tradition whose practitioners farmed, hunted, fished, and gathered across most of present-day Utah between two thousand and seven hundred years ago. All scholars of the Fremont from Morss to Jennings noted differences in material goods, architecture, and diet across the region. But they also believed that the commonalities they saw justified subsuming these peoples under the rubric Fremont. Some cultural differences certainly arose from the contrasting landscapes people lived in. Wetlands lie along the northern rim of the eastern Great Basin; arid and dissected canyons and plateaus dominate the land to the southeast; the Uinta Basin has a colder climate; and fertile, well-watered soils are present along the Wasatch Front. Each of these landscapes offered certain advantages and posed certain problems for Fremont people in making a living and creating the things of their world. Other differences stemmed from unique local historical trajectories, including interactions with neighbors. Fremont farming and village life was clearly an expression of the general Southwest farming tradition. But the Fremont people developed their own distinctive basketry, rock art, clay figurines, pottery, and ornaments.

Research on the Fremont people continues today, addressing old questions and raising new ones. In addition to environmental and historical differences among recognizably Fremont groups, researchers have found less visible sites away from larger villages and questioned whether some Fremont groups were farmers at all. When retreating floodwaters in the Great Salt Lake wetlands during the 1980s revealed ancient burials, analyses of them yielded unexpected insights into dietary diversity. This information led to a new understanding of the complexity of Fremont strategies for producing and gathering food.

Brigham Young University archaeologists excavated in Clear Creek Canyon in central Utah and at Baker Village on the Utah-Nevada border in the 1980s and early 1990s. Their work aroused renewed interest in Fremont communities and social life. More recent research in the Escalante River drainage has explored issues of ethnicity in the dynamic borderland between Fremont and ancestral Pueblo peoples.

**Fremont Origins**

What has all this research told us about Fremont people? The genesis of Fremont society began with influences and most likely migrants from the Southwest at least two thousand years ago. Inklings of a shift from reliance on hunting and gathering wild foods to farming come from several deeply buried sites, including the Elsinore Burial in the Sevier River valley, the Steinaker Gap site in the Uinta Basin, and the Confluence site on Muddy Creek in central Utah. House floors at the Confluence and Steinaker Gap sites were shallow ovals, some with central fire pits. Deep, bell-shaped storage pits containing corn lay adjacent to the houses. At Steinaker Gap, researchers found irrigation ditches and a check dam, a clear sign of the importance of farming. They found no pottery, but both sites had evidence of bows and arrows. Archaeologists working at the Steinaker and Elsinore sites discovered burials in the bell-shaped pits. In the former site, they found infants swathed in bone and shell beads, demonstrating how highly these communities valued children.

These findings tell us that by 500 CE, people were farming on the floodplains across much of present-day Utah. They lived near their irrigated fields, stored corn in deep pits, and used bows and arrows for weapons. Perhaps a hundred years later, they began making plain gray pottery. These practices sharply distinguished these people from foragers elsewhere in the Great Basin.

The Fremont culture reached its greatest geographical extent about 1050 CE. At that time, Fremont settlements ranged from present-day Brigham City on the north to Cedar City on the south. Many Fremont communities existed in the Uinta Basin and along the eastern flanks of the Wasatch Plateau, but they tended to be smaller than those found elsewhere. In all cases, Fremont people situated their villages near arable, well-drained land and water with which to irrigate crops. Many modern Utah towns, originally based on irrigation agriculture, overlie these old villages. Later sites along the Wasatch Front tend to be larger and located on ridges and knolls. More ephemeral sites are common, too. Some people camped at high elevations in the Uinta Mountains and on the Wasatch Plateau, and others set up temporary lodgings in the arid lowlands of western Utah. There they collected seasonal foodstuffs and critical resources such as tool stone.

**Shelter**

The Fremont people invested in architecture in keeping with their intentions: long-term residence required greater investment than temporary stays. With the materials available, they built pit houses and adobe-walled storage facilities along the Wasatch Front and at the Baker Village and Garrison sites on the Utah-Nevada border. They excavated pit structures up to three feet deep and twelve to fifteen feet across, usually making them circular in plan with vertical walls and flat bottoms. They roofed the houses with timbers reaching from the perimeter of the pit toward the center, covered the larger beams with smaller thatch or branches,

Figure 12.2. An adobe-walled granary with stone paving, Nawthis Village, central Utah.

and sealed it all with as much as a foot of dirt. In most cases, they climbed in and out of these houses via a ladder through a central roof hole. Shallow, roofed ventilator tunnels typically reached ten to fifteen feet south from the house perimeter, facilitating air circulation to move smoke out through the center entry. Sometimes residents plastered the pit house walls to hold rocks and dirt at bay. House floors were hard-packed dirt, perhaps cushioned with mats. A circular hearth with a clay rim in the center of the floor provided warmth and a place for cooking.

The Fremont also built large surface residences and granaries made of adobe blocks, each roughly a foot square and half a foot thick, laid like bricks to construct walls up to several feet high (no evidence remains of what completed the walls above that height). The best example of an adobe house and its associated granaries is at Nawthis Village in central Utah, but others exist at Baker Village, Nephi Mounds, and Bradshaw Mounds near Beaver, Utah.

Some researchers consider these to be more than residences—perhaps community gathering places.

Eastern Utah houses varied considerably. Some were shallow and edged with large basalt boulders. In Nine Mile Canyon, Fremont people built both storage structures and living quarters of stone masonry laid with mud mortar. Pit houses in the Escalante River drainage reflect influence from ancestral Pueblo neighbors. Similarities to Pueblo architecture include large flat slabs set vertically against walls to line pit houses and entry ramps and the defining of hearths and wing walls with stones set into adobe.

For temporary or seasonal use, Fremont people built brush houses. First, they cleared and smoothed an area and edged it with large rocks to support a tipi-like superstructure of poles joined in the center. They covered this frame with brush or perhaps hides. Such houses, referred to as wickiups, are most common in relatively remote areas and represent special-use camps associated with farming

Figure 12.3. A slab-lined Fremont pit house with wing walls at the Dos Casas site in the Escalante Valley.

villages, although some might have been residences of people who farmed little or not at all but whose material remains (ceramics and other tools) were Fremont in style. Archaeologists find wickiups in the Utah western desert and at high altitudes. These ephemeral structures might actually have been more common than the labor-intensive pit houses, but evidence of wickiups is less visible than remains of the more elaborate pit and surface houses.

The Fremont and other peoples who farmed in the Great Basin needed storage structures and granaries to store food, tools, and raw materials for future use. Early on, from about 100 to 400 CE, Fremont people kept such goods in bell-shaped pits. Later, as in Range Creek Canyon, they tucked granaries built of stone, adobe, and timbers into protected niches in cliffs, often well away from residences. Such a strategy of remote storage is likely evidence that the builders often spent time away from that region. The remote storage strategy contrasts with the adobe-walled granaries of one or two rooms adjacent to their pit houses. This adjacent storage strategy is found in the larger, more permanently occupied villages such as those found along the Wasatch Front.

**Making a Living**

What went into Fremont people's meals depended on the farming potential of the locale and the availability of wild foods. Cultivated crops such as corn, squash, and beans were important for everyone. Even those who might not have farmed had access to these foods through trade or perhaps raiding. Archaeologists find burned corncobs and kernels in most sites, regardless of location. But how much corn people ate varied widely, as researchers learned from analyzing human bones from Fremont-age burials in the Great Salt Lake marshes.

Fremont people also ate meat, as attested by the abundant animal bones left in trash dumps. Deer and cottontail rabbit bones outnumber all others in most sites along the Wasatch Front; mountain sheep replace deer in the more arid valleys and canyons to the east. Marshes adjacent to the Great Salt Lake and

Utah Lake attracted waterfowl, muskrats, and beaver and supported vast stands of bulrushes and cattails. The opportunistic Fremont made good use of these and other wild resources. Thousands of trout, sucker, and chub bones come from sites around Utah Lake, and near the Bear River marshes, sites have yielded abundant waterfowl remains.

The Fremont people used bows to shoot their prey, and arrow points are plentiful in some sites. Rock art depicts groups of hunters driving mountain sheep and other large game into nets, showing that the hunt was a communal pursuit. Drives and nets no doubt served for smaller game, too, such as molting waterfowl and speedy black-tailed jackrabbits. Snares would have targeted ground squirrels, cottontails, and perhaps grouse. Bone harpoons found in sites alongside lakes and streams are evidence of one fishing technique. Basketry traps, perhaps bows and arrows, and even people's hands were other effective ways to harvest spawning fish.

Wild plants, especially nourishing pine nuts and small seeds from rice grass, goosefoot, bulrush, and cattail, were important in Fremont diets, as were tubers and roots. The people stored hard seeds and nuts and preserved meat and fish by drying them. In short, like the foragers who came before them, the Fremont people made the fullest possible use of the wild resources around them. But unlike their predecessors, they added cultivated plants to their diet.

## Crafts

Gray-ware ceramics are signature Fremont artifacts (plate 13). Earlier peoples did not make pottery, and Pueblo neighbors to the south decorated pots in different styles and produced contrasting forms. Fremont potters most often made plain, round-bottomed jars, often with a single handle attached to the rim, handled pitchers, and painted bowls. They burnished pot surfaces with rubbing stones and decorated some jars and pitchers with incised designs or finely executed corrugations. They also used a unique technique called coffee-bean appliqué to decorate vessels and clay figurines. Their painted vessels, almost always bowls, feature black designs on a gray or, in a few cases, white background produced by applying a slip—a thin layer of finely ground, liquid clay—over the vessel's surface. Sometimes Fremont potters applied red ochre or hematite generously to decorate vessels.

Looking at basketry also helps to define Fremont culture. Fremont basket makers created shallow trays and some bowls using whole or split willows or other kinds of slender, pliable wood as the "rod," or the horizontally coiled foundation. They wrapped the rod vertically with even more pliable, thin woody strips. A fiber bundle lay on top of the rod, leading archaeologists to call this the "one-rod-and-bundle" technique. People also made twined bulrush and juniper bark mats to cover the floors of houses and line storage pits. From the dense bones of deer and other large animals, they made whistles, needles, long daggerlike awls, harpoons, rubbing tools for working hides, and what were probably weaving tools and gaming pieces. The ubiquitous awls served to split woody foundation rods for basket making and perhaps to repair clothing and footwear. Antlers served as wedges and flakers in tool making.

One type of artifact is uniquely Fremont: unfired clay figurines. These engaging, stylized human forms vary considerably in size and decoration across the region. Relatively elaborate specimens come from the northern Colorado Plateau. The most famous of these are the Pilling figurines, from a side canyon of Range Creek. Found in 1950 by Clarence Pilling of Price, Utah, these eleven figurines are notable for their large size (they average about six inches long) and elaborate decorations—necklaces, hair bobs, waist trappings, and red, buff, and brown face and body painting. Equally elaborate examples include a figure from the Nine Mile Canyon area that sports a face painted brilliantly in yellow-gold with vertical red stripes and a diagonal red "sash" across the torso. Noel Morss found many figurines in the Fremont River sites he explored, but most were simple, formed of dark red clay, with rounded heads and pinched noses.

In the early twentieth century, Charles W. Lee, a collector exploring sites in the Capitol Reef area, discovered a spectacular figurine wrapped in tiny blankets of animal skin, juniper bark, and cotton cloth and resting in a finely crafted toy cradleboard.

Other figurines featured skirts, elaborate hairstyles, earrings, and necklaces, giving us a glimpse of how the Fremont people clothed and adorned themselves. Actual items of clothing are surprisingly rare in Fremont sites, and we have to extrapolate what people wore from their figurines and rock art and from remnants found in burials. Kilts or skirts, occasionally decorated with bird or animal skins, may have been common. Rock art panels show figures with belts and possibly skirts, perhaps for both men and women. Another unique item associated with the Fremont is a type of moccasin that Morss found in the dry shelters of southern Utah. It was made from the lower leg skins of mountain sheep and formed in such a way that the dew claws or hocks were on the sole.

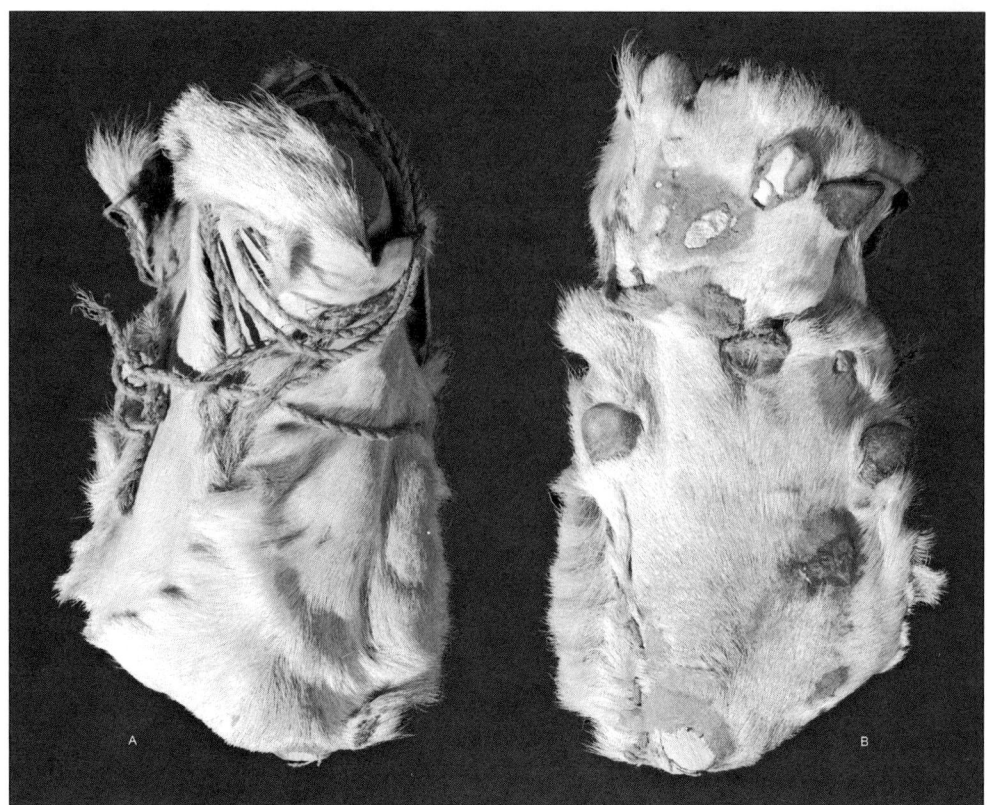

Figure 12.4. Top and bottom views of a pair of Fremont moccasins made from the hocks of a mountain sheep, Capitol Reef area, south-central Utah.

Although we know little about Fremont people's clothing, we know a lot about their accessories. They adorned themselves with stone and bone beads and pendants. Archaeologists have found trapezoidal bone beads that look much like elements seen on rock art panels and in miniature on figurines. At Nawthis Village, careful excavations exposed a necklace consisting of hundreds of finely made bone and black lignite disk beads. Such beads turn up in Fremont sites from Utah Valley to Escalante Valley and seem to have been the preferred form. We also find larger stone beads with striking color patterns and a few exotic beads made of marine shells (mostly Olivella shells) and turquoise. Neither occurs locally, so their presence shows that the Fremont traded with their neighbors to the south and west.

Besides jewelry, at least some Fremont people wore headdresses. Of the two that researchers have discovered, one is from Mantles Cave, in Dinosaur National Monument in northeastern Utah, and the other is from the Canyonlands area. The Mantles Cave example displays orange feathers from the red-shafted flicker, a bird common to the region. The feathers are arrayed vertically along a headband, making for a colorful display. The second example is made of a mountain sheep scalp with horn sheaths intact and decorated with Olivella shells. The considerable differences between these amazing specimens suggest that they served different cultural functions; possibly, they even reflect the importance of individual choice in Fremont society.

## Social and Ceremonial Life

We have few insights into Fremont social life. Indirect evidence for communal gatherings might be the many small, polished and decorated bone objects thought to be gaming pieces. These have

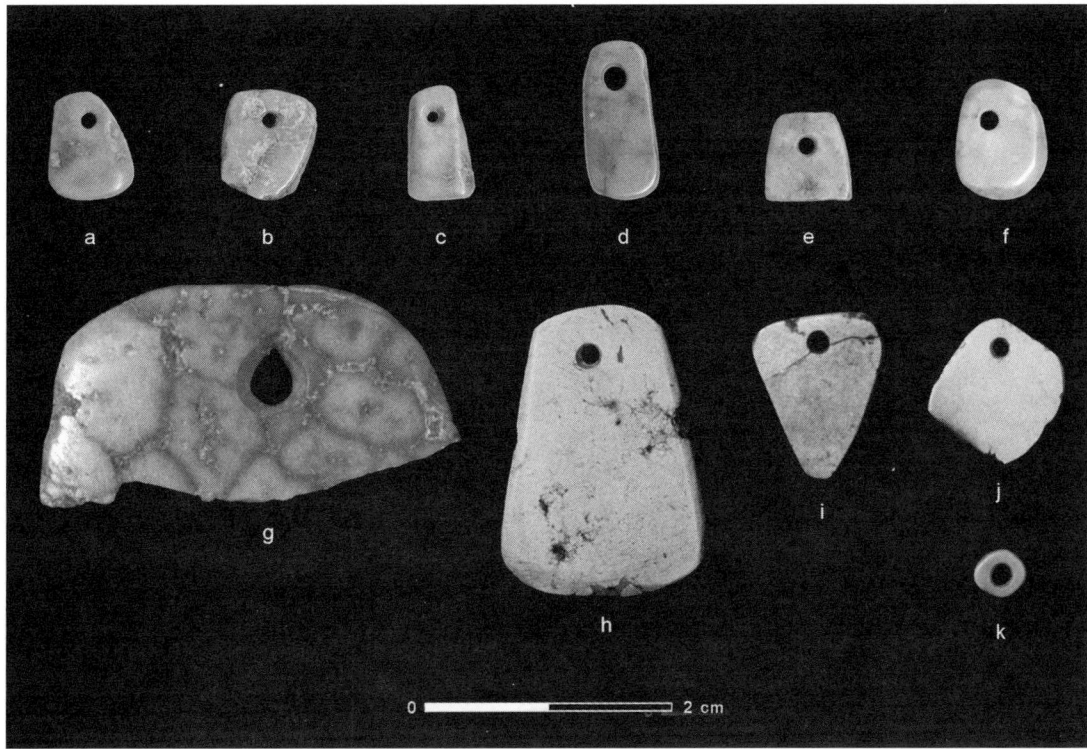

Figure 12.5. Turquoise pendants from Baker Village (*top row*) and Five Fingers Ridge (*bottom row*).

been collected in large numbers at some sites—hundreds from the Parowan Valley, for example. Gambling was a popular pastime in many Native North American societies, especially during festivals that brought people together for social and ritual activities. It makes sense to think that the bone pieces from Parowan Valley sites and elsewhere were used similarly during community gatherings.

Annual and semiannual festivals are well documented in the historic-period Great Basin and Southwest. These were important social events that brought people together for courting, trading, gambling, competing, and feasting. Evidence for lavish communal meals appears at the Baker Village site, where excavators found thousands of rabbit bones in a large, adobe-walled central structure. Whether they were the result of a single feast or of many, we are unsure. The rabbits themselves might have been captured in a communal hunt.

Festivals also provided opportunities for trading. Archaeologists have found marine shell, exotic ceramics, obsidian, and locally unavailable minerals in Fremont sites. For instance, researchers found more than a hundred Olivella shell beads and a dozen turquoise pendants at Baker Village. The shells came from the Pacific coast. The turquoise sources may have been in Nevada to the west or south. Turquoise from the Five Finger Ridge site, on the other hand, looks more like the turquoise beads and pendants found in Arizona and New Mexico.

Fremont people traded pottery across the region as well. Black-on-white painted bowls most likely made in Castle Valley appear in many sites well away from the Fremont area. Similarly, corrugated and black-on-gray painted wares apparently made in the Parowan Valley found their way north and west. Obsidian, the volcanic glass highly prized for making sharp tools, came primarily from sources along the eastern margin of the Great Basin, and obsidian tools are common finds in that part of the Fremont area. Obsidian is less common at Fremont sites on the Colorado Plateau, but the specimens found are from the eastern Great Basin sources. Even though we can see that Fremont people traded, we do not clearly understand the nature of that trade. Tool makers might have traveled to obsidian quarries or obtained blanks through down-the-line trade. Marine shell and turquoise probably passed from hand to hand along trade networks. Transporting pottery vessels would have been a challenge, and we still wonder how people moved them over such distances.

Not all social interaction, unfortunately, was friendly. Rock art panels portraying shield-bearing figures, some holding what may be trophy heads, evince conflict among the Fremont or between them and their neighbors. Charred, broken, and sometimes chopped human bones in houses raise the possibility of desecration or even cannibalism of enemies. The Turner-Look site yielded skull fragments, a possible "trophy" mandible, and the burial of a man who died from a blow to the head. Excavators of the Sky Aerie Charnel House site in northwest Colorado discovered a clay-capped hearth containing three human crania and other human bones. The excavators speculated that the heads had been roasted for consumption. Additional human bones lay scattered throughout the site, some near hearths and mixed with animal bones. These finds demonstrate that strife, even warfare, broke out on occasion during Fremont times.

On a more peaceful note, we find some evidence for community planning. At the Five Finger Ridge site in Clear Creek Canyon, the ancient residents positioned their adobe granaries along ridge lines and located most pit houses on slopes just below the ridge line. Perhaps this arrangement reflects coordination rather than simply personal preference. At Baker Village, Fremont people arranged both granaries and pit houses in a consistent way relative to the cardinal directions. This suggests directed group activity and implies the presence of a community head, an important indication of Fremont social and political structure. Researchers found an unusually large surface house there that might have served as home for a person or family of some importance, perhaps indicating some ranking in Fremont society.

The way people treat their dead can tell much about their social hierarchy or lack of one. The Fremont placed their deceased in pits inside or near abandoned houses. Unlike some other peoples, they did not consistently orient bodies in a certain direction, and some of the bodies were flexed, or bent, whereas others were buried fully extended. Nor did Fremont people normally bury many mortuary goods with the dead—the usual marker of high status—although they did sometimes place grinding stones in the grave. Few excavated Fremont burials have contained goods such as ornaments and ceramic vessels. The sparse offerings in most Fremont burials stand in stark contrast to ancestral Pueblo mortuary goods, which typically included ceramic bowls and jars, beads, and other ornaments.

Exceptions to this burial pattern include two burials of adult men in the Parowan Valley that contained bone tools and evidence of bird and weasel skins, perhaps clothing decorations. A burial found in Huntington Canyon, also an adult man, lay beneath a house floor in a pit sealed with several grinding slabs. Two associated caches contained arrow points and miscellaneous chipped stone and bone tools. Many objects lay on a bench encircling the floor, among them, several unfired but painted clay figurines, two unusual ceramic vessels shaped like cornucopias, a miniature clay cradleboard, red and yellow pigments, textiles, corncobs, and beans and other food items. Another atypical burial site is a possible mass grave containing as many as seven persons, uncovered in the Great Salt Lake marshes. There the grave goods included ground stone, bone awls, gaming pieces, needlelike tools, and chipped stone items smeared with red ochre.

What status did these people with more elaborate grave goods hold in their communities? We do not know, but these remarkable collections of items, especially those with the Huntington burial, imply some kind of special status. Perhaps the deceased were shamans, respected leaders, or successful hunters.

**Rock Art**

Rock art is the remnant of Fremont presence that is easiest to see. Fremont artists left their marks on many stone canvases—sandstone cliffs, alcoves, and boulders. Panels attributed to the Fremont are among the best known in the world. Painted and pecked geometric, human, and animal shapes decorate these stone panels. Insightfully composed hunting scenes vie with hodgepodges of elements accumulated over time. Most resist easy understanding. Scholars Sally Cole and Polly Schaafsma have ably described the Fremont style of rock art.

Figure 12.6. Imposing classic Fremont figures wearing heavy necklaces, Northeast Utah. One holds a head trophy.

Anthropomorphs—broadshouldered, trapezoidal or hourglass-shaped human figures—are ubiquitous. Many of them wear elaborate headdresses and display body ornaments.

Despite shared features, the Fremont style varied geographically. West of the Wasatch Range, rock artists put greater emphasis on geometric textile or pottery designs, whereas on the Colorado Plateau, they more often depicted anthropomorphs and developed the form more highly. Heroic figures wearing elaborate headgear and holding bags, trophies, or weapons intimidate viewers today and may have in the past. Multiple humpbacked (or backpacking?) figures drawn in single file might depict the way goods were moved from place to place. Artists portrayed mountain sheep, elk, and other animals in ways that reflect intimate knowledge of animal behavior, as well as means of capture.

Understanding the function and meaning of rock art remains difficult. Scholars speculate that the images might represent hunting or shamanistic rituals, or perhaps they served to mark territory or record a concern with warfare. The frequency with which the panels portray large game, especially mountain sheep but also elk and deer, supports the notion that rock art was meant to convey information about animal behavior and perhaps success or failure in actual hunts. Polly Schaafsma has suggested that the elaborate headgear and attention to details of dress and associated paraphernalia in rock art indicate ceremonial activities. That some figures in Dinosaur National Monument and Capitol Reef National Park wear feathered headdresses similar to the piece discovered at Mantles Cave might mean that some rock art depicted actual persons. Although rock art panels exist throughout the Fremont area, certain places are treasure houses of Fremont rock art: among others, Clear Creek Canyon in central Utah (plate 4), several places in the Uinta Basin, Nine Mile Canyon, Range Creek, and Capitol Reef National Park.

**The End of an Era**

Researchers find few Fremont sites dating after 1300 CE, but changes in this distinctive way of life began much earlier. People living in the Great Salt Lake marshes ceased to eat corn at about 1150, although those dwelling in the Salt Lake Valley to the south continued to consume this staple into the 1200s. Pit house dates from Five Finger Ridge bunch up between 1200 and 1300, but none falls after 1350. We find some evidence of continued farming as late as the 1400s in the northeastern portion of Fremont territory, but clearly, people were giving up the practice in many areas. Reasons for the decline of farming are complex, but an important one may have been that shifts in summer rainfall made farming difficult and forced some people to emigrate to places where growing corn was still possible. Hunting and gathering continued as they had for millennia, but the Fremont style of combining these practices with farming disappeared. This shift might signal the arrival of the region's historic peoples, the Utes, Southern Paiutes, and Shoshones.

**Joel C. Janetski** is emeritus professor of anthropology at Brigham Young University. His views on the Fremont are based on twenty-five years of archaeological research in the eastern Great Basin and Colorado Plateau.

Figure 13.1. Hohokam faces modeled in clay between about 900 and 1150 CE.

# The Hohokam Millennium

*Suzanne K. Fish and Paul R. Fish*

Hot, dry regions of the world have produced some of the most memorable preindustrial civilizations, and the southern deserts of Arizona are no exception. The aptly named modern Phoenix, now the fifth largest city in the United States, arose not from the ashes but from the ruins of what was the most populous and agriculturally productive valley in the West before 1500 CE. When the early Southwestern archaeologist Frank Hamilton Cushing entered this Salt River Valley in 1892, he climbed atop an earthen monument in what would become urban Phoenix and exclaimed at the discovery, "One of the most extensive ancient settlements we had yet seen.... Before us, toward the north, east, and south, a long series of...house mounds, lay stretched out in seemingly endless succession." Entrepreneurs arriving from the eastern United States a few decades earlier had, like Cushing, seen not only house mounds but also the former courses of the most massive canals ever built in the pre-Columbian Americas north of Peru. They soon reestablished large-scale irrigation by laying out new canals virtually in the footprints of the prehistoric ones, triggering the growth of the future city.

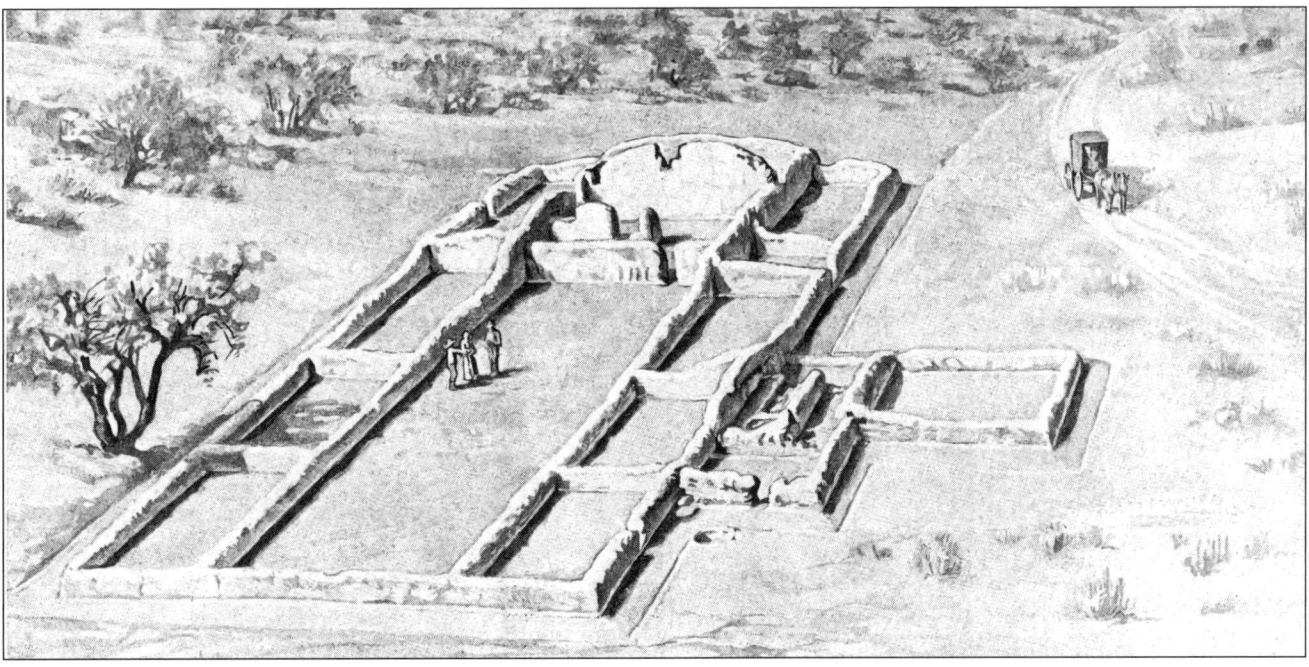

Figure 13.2. Centuries of weathering reduced Hohokam adobe buildings to low "house mounds" of earth. When excavated, the mounds often reveal well-preserved outlines of walls, as in this compound at Casa Grande National Monument excavated in 1908.

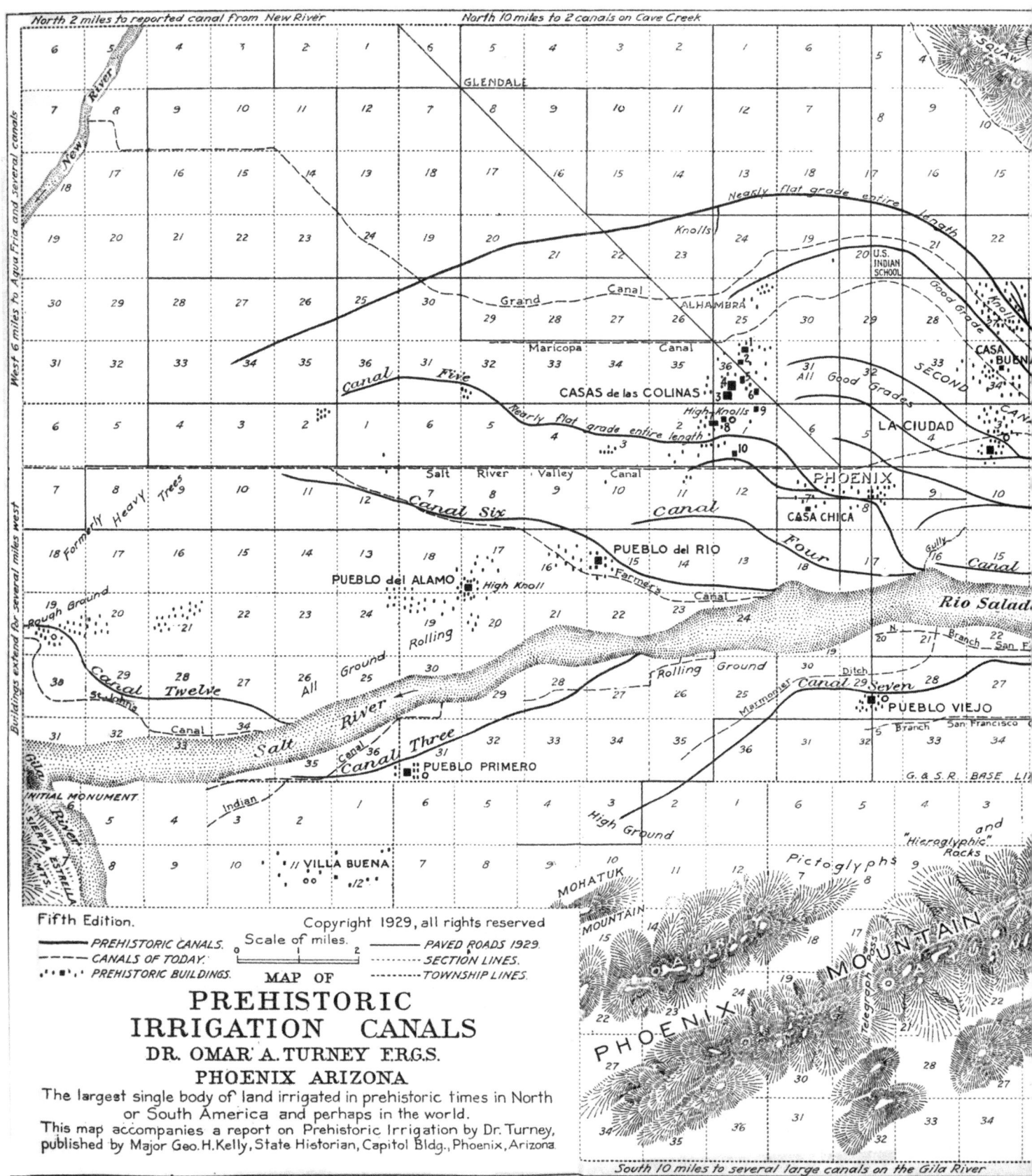

Figure 13.3. Omar Turney, engineer for the city of Phoenix, compiled this map of major Hohokam sites and canal systems in

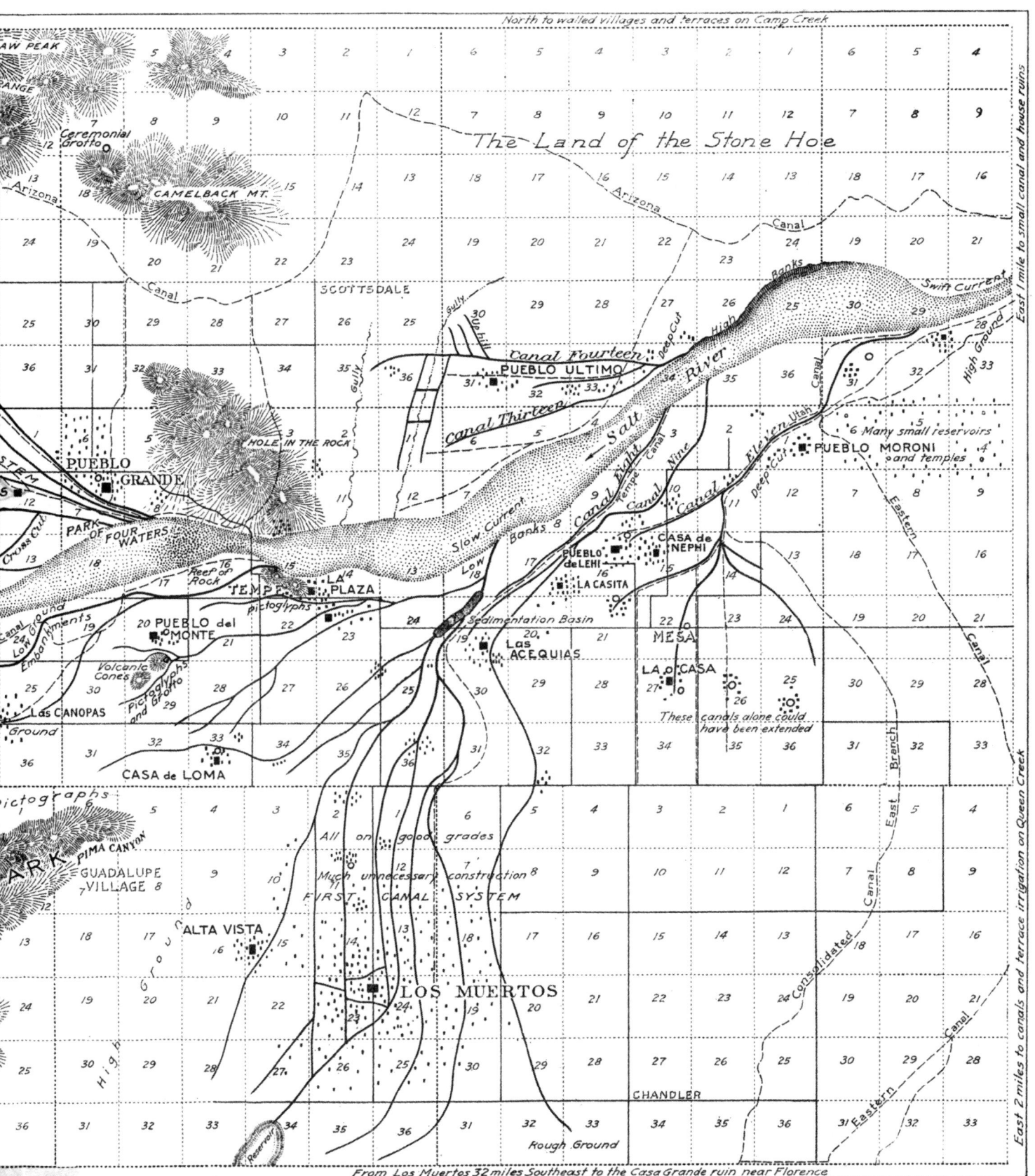

the 1920s, on the basis of earlier records and remains still visible at the time.

Figure 13.4. Shells with etched designs are limited to the Hohokam Sedentary period (900–1150 CE).

The remarkable people whom archaeologists call the Hohokam were the builders of the earthen monuments, adobe houses in profusion, and huge canals that so impressed later visitors to the Salt River Valley. From 450 to 1450 CE—the "Hohokam millennium"—the basin at the confluence of the Salt and Gila Rivers formed the core of their geographic and cultural domain. For a thousand years, the Hohokam maintained a recognizable cultural identity among the diverse peoples who inhabited other parts of the prehistoric Southwest and adjacent northwestern Mexico.

## Who Were the Hohokam?

The fragments of buff to brown pottery with red painted designs that litter the low-lying basin floors of southern Arizona are the most distinctive and abundant material remains of former Hohokam residents. Ingenious farmers who employed an assortment of agricultural strategies to grow crops in arid terrain, they ultimately engineered irrigation networks surpassed in length and size only by the canals of Andean empires. In addition to creating unique artifact styles, the Hohokam set themselves apart from the ancestral Pueblo, Mogollon, and other archaeological cultures of the Southwest by the forms of the public buildings in their largest villages. These ballcourts and platform mounds reflect the characteristic beliefs and community rituals of the Hohokam.

What might it have meant to individuals, household members, and villagers to be participants in the Hohokam cultural sphere? It is difficult to answer this question from the fragments that have survived for archaeologists to examine. Yet, the fact that they shared the same ways of making and decorating pottery, as well as other canons of style and utilitarian design, tells us that they were in close communication with one another and held common understandings about such matters. That they shared crops and farming technologies shows that they turned to the same solutions to meet the challenges of desert cropping. That they built the same sorts of structures for communal rituals implies that a shared set of beliefs guided them. But archaeologists cannot determine whether all the ancient Arizonans they classify as Hohokam spoke the same language or whether they considered themselves to be members of the same ethnic group or culture.

Figure 13.5. Partially excavated ballcourt at Snaketown. The earthen banks of ballcourts enclosed the playing field and provided a vantage for spectators during ball games or other public events.

Why these uncertainties over the meaning of being Hohokam? First, the distinctive archaeological remains that identify the Hohokam heartland are spread over an expanse of almost thirty thousand square miles in the southern half of Arizona, an area larger than the state of South Carolina. The hallmarks of Hohokam culture are generally bounded by the upper reaches of the Agua Fria and Verde Rivers to the north, the Mogollon Rim to the northeast, the Dragoon Mountains to the southeast, the Mexican border to the south, and the Growler Mountains to the west.

Within this far-flung territory, archaeological remains have much in common, but they also vary in important ways. Inhabitants of some sectors chose only parts of the overall cultural package to incorporate into their lives. For example, in the Tonto Basin, on the northeastern edges of the Hohokam domain, local people using red-on-buff pottery never built ballcourts, but they eventually erected platform mounds. Migrations of Hohokam and non-Hohokam groups into the Tonto Basin contributed to the mixing of cultural practices. Where local groups shifted between full and incidental participation in Hohokam cultural traditions at different times, the archaeological boundaries for the Hohokam shift accordingly.

A second reason for our uncertainties is the area's historic ethnic diversity. When Spanish explorers arrived in the late seventeenth century, they found Native Americans with diverse languages and life styles all living in the former Hohokam domain. This included groups speaking primarily Piman languages (O'odham dialects) in the central portion, people speaking Yuman languages (Colorado River Yuman to the west and Yavapai to the north), and groups speaking Athabascan languages (Western Apache) in the northern and eastern reaches. The diversity of the postcontact era suggests that the Hohokam, too, might not have been homogeneous in all respects. It also complicates the question of how the prehistoric Hohokam are related to the succeeding native occupants of the same region.

Figure 13.6. The Hohokam of the Preclassic period used carved stone palettes in household and public rituals.

## How Are the Hohokam Remarkable?

Among preindustrial societies throughout the world, the Hohokam hold the distinction of having constructed massive canal networks (up to twenty-two miles in length) and irrigated extensive tracts of land (up to seventy thousand acres) in the absence of state-level government and a corresponding level of societal complexity. Archaeologists have not yet identified the graves or dwellings of rulers with such obvious high status and power that they could have imperiously resolved the inevitable disputes that arise among multitudes of water users or regulated the huge labor force needed to build and maintain the canals. Nor have archaeologists found evidence of a developed Hohokam bureaucracy that could have provisioned and organized workers. Yet, the canal systems alone clearly required a tremendous amount of coordinated labor. Jerry Howard, an expert on Hohokam irrigation, estimates that it would have taken nearly a million person-days of labor to construct the trunk lines of just one of the Phoenix Basin canal systems. That figure does not include the additional effort needed to build secondary lines out to fields, clean out annual buildups of canal sediments, and make repairs after storms and floods.

The Hohokam also constructed earthen ballcourts and platform mounds of modestly monumental size relative to those found elsewhere in the ancient world, again without all-powerful rulers or an established bureaucracy. The placement of these monuments imparted a unique pattern to Hohokam landscapes. Large villages with ballcourts or platform mounds appear about every three miles along major canal lines in the Phoenix Basin and at greater intervals among surrounding settlements. The largest villages stood at the centers of clusters of smaller settlements, each cluster forming an organizational unit of population and territory that Hohokam archaeologists call a "community." The monuments in the centers served as staging areas for communal events unduplicated in outlying settlement zones. This characteristic mode

Figure 13.7. Saguaro cholla, prickly pear, and small trees with edible beans supplied the Hohokam with important foods in the Sonoran Desert.

of community organization both accommodated and shaped Hohokam economic, political, and ritual life.

Ballcourts and platform mounds are unusual in the US Southwest in their resemblance to the monumental forms of Mesoamerica, the heartland of the Toltec, Aztec, Maya, and other high cultures centered in what today is Mexico. Hohokam stylistic motifs and artifacts that are related to ritual and ideology, such as figurines, palettes, and censers, also show a pronounced Mesoamerican inspiration. Many questions about the nature of this cultural connection linger unanswered because archaeologists until recently have mostly neglected the four hundred miles of northwestern Mexico separating the Hohokam from the most likely west Mexican sources of such Mesoamerican traditions. A stronger Mexican connection than is seen elsewhere in the Southwest is further apparent in the Hohokam trade for copper bells, iron pyrite mirrors, marine shells to make into jewelry, and a few other items that originated south of today's border.

The Hohokam are especially notable for the long-term continuity of their lifeways. In comparison with peoples in other parts of the Southwest, the Hohokam tended toward unusually prolonged residence in place. Once established, some clusters of dwellings in the largest settlements persisted—renovated, extended, and rebuilt—up to several hundred years. Central plazas in these foremost settlements remained the heart of village life from beginning to end. Successive generations lived in many of the largest settlements, amid irrigated land, for more than half the Hohokam millennium, and farming families returned again and again to outlying settlements where crops could be watered by

Figure 13.8. A street vendor in Guaymas, Sonora, sells sweet slices of baked agave hearts.

alternative means. Settlement stability was an outcome of the productivity and sustainability of Hohokam agriculture. Sustainable production in turn was closely tied to places where enough water could be predictably captured and delivered to crops.

## The Sonoran Desert Environment of the Hohokam

The great majority of Hohokam people lived within the outlines of the Sonoran Desert in southern Arizona and within the range of the towering saguaro cactus, one of its distinguishing species. Sonoran Desert vegetation differs from that of the Chihuahuan Desert to the east and the Mohave Desert to the west, thanks to rainfall that arrives in both winter and summer rather than mostly at one time of year. The two seasons of rainfall allow the Sonoran Desert to support large cacti such as saguaro (plate 10) and cholla and dryland trees such as mesquite, ironwood, and paloverde, in addition to the shrubs common to all three deserts. The fruits and buds of the cacti and the beanlike pods of the trees provided plentiful and reliable wild staples in the Hohokam diet. Groves of mesquites and plants with edible small seeds, including saltbush, grasses, pigweed, and goosefoot, flourish along Sonoran Desert watercourses. For most of the meat they consumed, the Hohokam hunted jackrabbits, cottontails, and other small animals on land surrounding their homes and fields. As the human population increased, hunters had to go farther afield for large game, pursuing deer and bighorn sheep at higher elevations. The wild resources of the Sonoran Desert added variety, nutritional balance, and back-up supplies in times of poor harvests.

Hohokam everywhere experienced the risks and opportunities of their Basin-and-Range environment. They focused their day-to-day lives as farmers on land in the basin interiors. They seldom lived in the mountains at basin edges and only occasionally sought out upland resources. Temperatures typically topped 100° F on ninety days or more per year, and annual rainfall varied from seven to fifteen inches. The vast highland watersheds of the Salt and Gila Rivers allowed the Phoenix Basin Hohokam to fill miles of canals. Farmers in other basins used floodwaters in tributary streams after heavy summer rains, along with smaller-scale canals, to water their crops. The Hohokam raised corn, beans, squash, and cotton in irrigated and floodwater fields. They also trapped surface runoff in stone grids, on low terraces, behind checkdams, and under mulches of piled rock on dry slopes to grow smaller amounts of these crops and to raise agaves for food and fiber.

**Hohokam Historical Trajectories**

The beginnings of agriculture in Hohokam country at about 2000 BCE kicked off a rise in population and an increase in societal complexity that would span the Hohokam millennium. The arrival of domesticated corn, or maize, from Mexico curtailed the seasonal movements of the hunters and gatherers who had populated the Sonoran Desert before this time. By 1500 BCE, early cultivators in the Tucson Basin were constructing irrigation ditches in small settlements along the Santa Cruz River. Archaeologists find many large food-storage pits in and around the small, circular houses of these early farmers. Along with the substantial labor invested in building canals and maintaining fields, stored harvests suggest that people stayed in their settlements for much of the year.

An important transition in the organizational scale of society about 450 CE coincided with the consolidation of patterns in artifact styles, architecture, and economics that archaeologists define as Hohokam culture. People came together in more permanent settlements with well-built pithouses. Homes surrounded central plazas in the largest villages. Soon, Hohokam people in the Phoenix Basin began to construct the massive irrigation systems for which they are famous. Hallmarks of Hohokam culture such as ballcourts, red-on-buff pottery, palettes, and censers made their first appearance, and people began to cremate their dead, a practice common among the Hohokam. Ritual objects and ballcourts signaling participation in Hohokam ideology reached their greatest regional extent between 700 and 1150 CE, a time span that archaeologists call the Hohokam Preclassic period. During the same interval, cultural developments in Chaco Canyon peaked, and Chacoan-style "outlier" settlements proliferated across the Pueblo Southwest to the north of the Hohokam.

The transition to the Classic period after 1150 CE marked a watershed in Hohokam culture. Phoenix Basin potters produced less and less of the trademark red-on-buff pottery and eventually stopped making it entirely in favor of pan-Southwestern styles. Rather than continue to arrange pithouses in small groups around a shared courtyard, villagers began to build larger groups of adobe rooms inside walled compounds. Toward the end of the Preclassic period, the Hohokam stopped building and using ballcourts. Instead, as the Classic period opened, they began erecting platform mounds with rooms on top. Like the new adobe houses, the mounds were enclosed within a wall. They reflected a new set of rituals and beliefs that included the acceptance of a growing hierarchy among social groups. Local inhabitants built platform mounds in an area smaller than that over which ballcourts had once been distributed. Canal systems in the Phoenix Basin, in contrast, reached their greatest extent, and cultivation away from the rivers increased. Most Hohokam subareas reached their maximum population during the Classic period, while population densities increased at the largest centers. Pueblo people migrated from the north into the Hohokam basins, heightening the diversity of the occupants.

Sometime between 1400 and 1550 CE, Hohokam society collapsed, and the Hohokam disappeared as a coherent archaeological culture. Because archaeologists have found so little evidence for what really happened at the end, they hold conflicting opinions and promote different scenarios. The long record of large and sustained agricultural settlements within Hohokam boundaries ends without any clear transition to groups with new cultures.

Figure 13.9. Emil Haury looks down an excavated canal at Snaketown in 1964.

We have little information about the people of the Phoenix Basin until the Spanish Jesuit missionary Father Kino visited the area more than a century later, in the 1680s. By that time, indigenous peoples did not closely resemble the Hohokam.

One current scenario, based on reconstructions of annual Salt River stream flow from tree-ring data, sees disastrous fourteenth-century floods leading to unpredictable harvests, hunger, and disease, forcing many people to leave the region. According to another view, an increasingly hierarchical and demanding leadership fostered political instability and was overthrown from within. O'odham oral traditions describe events of this sort. Other archaeologists propose that the deadly new diseases introduced into central Mexico by the Spaniards traveled rapidly along the trade routes, dealing a devastating final blow to the Hohokam.

## Hohokam Archaeology and Archaeologists

A few pioneering Southwestern archaeologists came to Arizona in the late 1800s and early 1900s to excavate at major Hohokam sites for patrons and institutions in the East, but they did not maintain these interests throughout their careers. The first person to dedicate himself to studying the ancient people of southern Arizona was Harold Gladwin, who established his own research station, called Gila Pueblo, and went about defining the extent of the "Red-on-Buff culture" in the early 1930s.

Lacking professional training as an archaeologist, he hired a young scholar named Emil W. Haury to assist in his excavations at Snaketown on the Gila River, the most influential of all Hohokam sites. The central importance of Emil Haury and his work to Hohokam archaeology cannot be overstated. During his long and distinguished career at the University of Arizona, Haury returned to Snaketown in the 1960s and later published a report that remains the classic reference for Hohokam studies.

A relatively small number of publications on the Hohokam appeared before the early 1980s, when a rapid change in the structure and personnel of Hohokam scholarship was just getting under way. A complex of new federal and state laws mandated that archaeological remains be inventoried and investigated before land could be developed. University faculty and students were joined by archaeologists in growing numbers of private companies that formed to meet the demands of Arizona's dramatic urban growth and large-scale federal land and water projects. Federal, state, county, and city agencies and, finally, tribal governments hired archaeologists and established programs to oversee threatened archaeological resources. In Arizona, the Bureau of Reclamation and the Arizona Department of Transportation were major funders of Hohokam research, sponsoring large projects and a continuing series of smaller efforts. An explosion of publications resulted in the following decades. In just twenty-five years, the Hohokam domain became one of the most intensively studied regions in the world, and scholars now must scramble to absorb the exponentially expanding archaeological data.

**Suzanne K. Fish** and **Paul R. Fish** are both curators of archaeology at the Arizona State Museum and professors of anthropology at the University of Arizona. Suzanne Fish's research in the Arizona-Sonora borderlands involves ethnobotany, traditional farming, archaeological settlement patterns, and Hohokam social and political organization. Paul Fish's Hohokam research emphasizes settlement patterns, farming systems, and the emergence of complexity.

Figure 14.1. Hopi-Tewa potter Nampeyo, around 1906. Hopi-Tewas descend from Tewa people who, around 1700, migrated from the northern Rio Grande region to join the First Mesa Hopi community, where they accepted the Hopi way of life.

# Pottery of the Sierra Sin Agua

*Kelley Hays-Gilpin and Christian E. Downum*

At first, locally made pottery of the Sierra Sin Agua is difficult to see as one strolls through the ruins of a Hisat'sinom village (plate 6). Small fragments of brown and red pots blend easily with the earth tones of clay soils. Next to showier black-on-white and red-on-orange pottery imported from neighboring regions, local jars and bowls at first glance seem plain and drab. But look more closely—this warm-hued pottery displays a pleasing range of mottled colors caused by the touch of fuel, air, and fire. Delicate pieces broken from finely polished bowls attest to the skill of potters who shaped subtle beauty and coaxed elegant function from materials readily at hand.

The earliest pottery in the Sierra Sin Agua, made around 600 CE, consisted of small, lightly polished brown vessels, embodying a simpler technology than those evolving elsewhere on the Colorado Plateau. Archaeologists classify this pottery as Alameda Brown Ware, and it endured, with slow changes and refinements, for as long as the Hisat'sinom lived around the San Francisco Peaks. Its makers, people whom archaeologists traditionally referred to as the Sinagua, lived on the southeastern side of the peaks. Like other Native peoples throughout the Southwest, they shaped clay into myriad forms: bowls, jars, ladles, pitchers, cooking pots, effigy vessels, spindle whorls (small weights that helped weavers spin cotton thread on a wooden spindle), animal and human figurines, beads, pendants, and pipes.

The makers of Alameda Brown Ware formed their pots from locally available volcanic clays, to which they added temper, a nonplastic material such as sand that helped prevent the clay from cracking as it dried. Tempering materials consisted of whatever was available nearby, reflecting the diverse geology of the Flagstaff area. Common tempers included volcanic ash, tuff, sanidine crystals, quartz or basalt sand, crushed basalt, and crushed potsherds. Potters made Alameda Brown Ware vessels by the "paddle and anvil" technique, first building up coils of clay and then shaping and thinning them with a wooden paddle, which they struck against a stone or ceramic anvil held on the inside of the pot.

Alameda Brown Ware is unusual in that although potters sometimes polished the interiors of bowls using smooth, water-rounded pebbles, they rarely painted or decoratively textured these vessels.

Figure 14.2. Ancient pottery-making tools from the Sierra Sin Agua. *Top:* anvils, the two on the left of limestone, the one on the right of basalt. *Center:* wooden paddle for smoothing clay coils against an anvil. *Bottom right:* polishing stones.

Figure 14.3. San Francisco Mountain Gray Ware jars and a mug (*front left*).

Instead, they left them smooth and plain. Sometimes the makers smoothed a red slip—a slurry of clay—over a vessel's exterior as it dried, finishing it by rubbing it to a light polish with a smooth stone.

Makers of Alameda Brown Ware fired their pots in what is called an "oxidizing" atmosphere—that is, in such a way that air flowed freely through the mound of firewood forming the "kiln." The behavior of iron-rich volcanic clays under such firing conditions, which could not be tightly controlled, produced subtle variations in color, from brown and red to black. Darkened patches known as fire clouds formed where fuel wood touched a vessel wall or came close enough to produce an oxygen-poor or carbon-rich microenvironment at the vessel's surface.

Two of the most common types of Alameda Brown Ware are what archaeologists have labeled Sunset Brown and Sunset Red. People probably carried and stored water in red-slipped Sunset Red jars, and they likely served food in Sunset Red bowls. Both types contain shiny black particles of volcanic ash from the Sunset Crater eruption. Because volcanic ash is hard and angular, ash-tempered pottery is especially strong and resists cracking from stresses such as the thermal shock that comes when a vessel is used for cooking.

Archaeologists named ancient people living on the northwest side of the San Francisco Peaks a separate culture, the Cohonina, partly because they made and used pottery different from that of their Sinagua neighbors. Now called San Francisco Mountain Gray Ware, this pottery is gray, thin, and grittier than Alameda Brown Ware. Cohonina potters usually selected iron-poor clays and added sand as temper. They, too, used the paddle and anvil technique to thin their vessels, but then potters scraped them on both sides with a shaped potsherd or piece of gourd rind, leaving parallel striations. They fired the pots in pits covered with wood and charcoal to create an oxygen-poor atmosphere. This "reducing" atmosphere generally turned the vessels gray, with splotches of black.

Cohonina potters made jars, pitchers, and

Figure 14.4. Tusayan White Ware vessels. The piece at bottom center is a bird effigy jar.

bowls in simple but elegant shapes. Often they left the surfaces plain, but sometimes the maker applied a thin, red clay wash after the pot was fired. Because this slip is prone to dissolving or wearing away, researchers call the type Deadmans Fugitive Red. Rarely, Cohonina potters painted bowl interiors with a plant-based paint. The designs resemble those from the Kayenta Pueblo area to the northeast, but Cohonina paint is often dark gray rather than true black and the designs are simpler. Apparently, Cohonina people were satisfied with the subtle contrast between dark paint and gray background; they never tried to increase the contrast by adding a white slip on which to paint the darker designs.

Differences between San Francisco Mountain Gray Ware and Alameda Brown Ware are greater than the immediately obvious ones of color and surface treatment. For example, pitchers make up about 20 percent of all known San Francisco Mountain Gray Ware vessels but are almost nonexistent in Alameda Brown Ware. Alameda Brown Ware pots tend to be wider and shorter than those of San Francisco Mountain Gray Ware. The height-to-diameter ratios of the wares form two distinct statistical populations, with little overlap. We suspect that the development of pottery territories coincided with a sense of cultural identity, although local identities were probably more complex and dynamic than the simplifying terms *Sinagua* and *Cohonina* imply.

The Hisat'sinom of the Sierra Sin Agua, self-reliant, made most of their own pottery. Yet, to greater or lesser extents, they also acted as consumers of pottery, importing vessels from their neighbors in all directions. Especially popular were

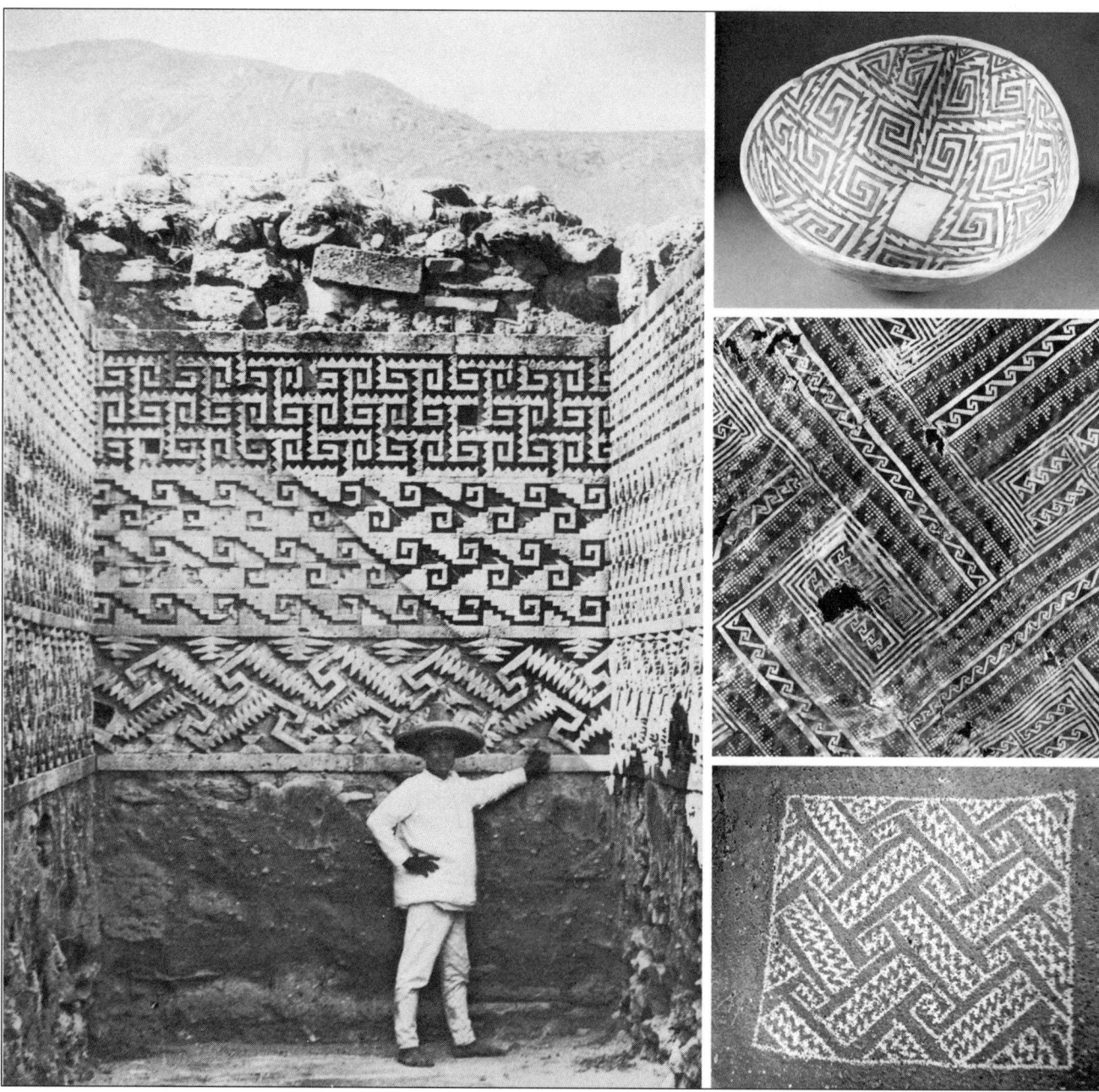

Figure 14.5. Mesoamerican design elements on pottery, textiles, and rock art from the Sierra Sin Agua and the Verde Valley, about 1140–1220 CE. *Left:* wall mosaics at the ancient site of Mitla in the Mexican state of Oaxaca. *Right, top to bottom:* Flagstaff Black-on-white bowl, about 1140–1220; painted cotton textile from Hidden House Ruin, Arizona, about 1200; petroglyph near Citadel Pueblo, Wupatki National Monument, about 1200.

the beautiful, intricately painted, black-on-white and black-on-red bowls and jars manufactured to the north and east.

Potters of the Kayenta tradition, ancestors of the modern Hopi people, made their wares mostly in the vicinity of Black Mesa, a massive plateau looming over the modern town of Kayenta, Arizona, across the Little Colorado River northeast of the Sierra Sin Agua. Neighboring peoples valued Kayenta pottery highly, and ancestral Hopi potters over time exported hundreds of thousands of their pots to outlying regions. At Hisat'sinom ruins in the

Figure 14.6. Tusayan Gray Ware vessels. The large jar at center and several of the small pots have "corrugated" surfaces.

Sierra Sin Agua, 5 to 25 percent of all pottery consists of Kayenta wares.

Imported pottery of the Kayenta tradition came in white, gray, and red or orange. Tusayan White Ware, consisting mostly of food serving bowls and water storage jars, was polished and then decorated with plant-based paint. When Kayenta potters fired their carbon-rich, iron-poor clay in a reducing atmosphere, they got white backgrounds that contrasted sharply with the overlying black paint. Their intricate designs generally took geometric shapes, with individual elements repeated and rotated in wide bands running across the outside of jars and the inside of bowls. Many of the design elements, which crosscut media such as ceramics, rock art, and textiles, can be traced ultimately to decorative traditions of Mesoamerica.

The gray Kayenta pottery that shows up in Sierra Sin Agua archaeological sites, Tusayan Gray Ware, was fashioned from the same clay and with the same techniques as Tusayan White Ware, but these cooking and storage jars had coarser sand temper and rougher surface textures. To decorate the surfaces, potters often pinched or indented the coils in a manner called "corrugation." The coarse temper and corrugations might have enhanced the cooking performance of these pots.

Two kinds of decorated red and orange pottery were also popular imports to the Sierra Sin Agua. People carried San Juan Red Ware, produced in southeastern Utah, far and wide perhaps as early as the 700s CE. Its most common type, Deadmans Black-on-red, is fine, thin walled, and hard, with a well-polished red slip and bold black designs. The painted designs usually include long parallel lines to which are attached interlocking solid shapes such as hooked triangles.

Sometime in the 1000s, San Juan Red Ware was replaced by Tsegi Orange Ware, whose makers used an iron-rich clay that turned bright orange when fired. By the 1100s, these potters had developed a three-color ("polychrome") style of decoration,

Figure 14.7. Little Colorado White Ware vessels.

featuring designs in red and black painted over the orange background. To modern eyes, it is easy to understand the appeal these colorful red and orange pots must have held for Hisat'sinom who bartered for them.

One last source of imported pottery was the Hopi Buttes area along the Little Colorado River, south of the Hopi Mesas. Potters there made black-on-white serving vessels and gray, corrugated utility jars. Archaeologist Amy Douglass used sophisticated chemical and geological techniques to match Little Colorado White Ware pottery to a clay that outcrops in the eroded badlands near modern Dilkon, Arizona. This iron-rich clay fires dark gray in a reducing atmosphere, and nothing a potter can do—no amount of polishing, no special firing techniques—will make it appear white. To make a white pot from this clay, the maker must cover it with a white slip before it is painted and fired. All Little Colorado White Wares exhibit a white, chalky clay slip, probably made from nearby sources of kaolin, a pure white clay.

One might expect that all these beautifully painted white, red, and orange imports—displaying some of the most complicated and skillful painting in all of ancient North America—would have inspired Sinagua and Cohonina potters to decorate their own brown ware pots. In fact, they seldom did so. They knew how to paint pottery and had access to the necessary pigments, but at most, they emulated the simplest designs of their Kayenta neighbors and never developed a locally distinct style. Sinagua potters occasionally painted a few white, broad-line geometric designs on vessel exteriors, imitating Hohokam pots for a brief time in the late 1000s and early 1100s. Cohonina artisans now and then painted relatively simple designs covering only a small fraction of the vessel.

Why did the Sinagua and Cohonina people never embrace their neighbors' pottery painting traditions? Perhaps they simply invested greater effort in media such as textiles, tattoos, or jewelry that for them more effectively communicated information about social identity. Perhaps they simply valued the distinctiveness and subtle attractions of the

### Endangered Pottery

Ancient pottery tells us a great deal about the daily lives and interactions of the people who made and used it. To the Hopi people, potsherds are footprints of their ancestors. At each place where the clans stopped during their migrations, they broke their old pots and made new ones so that those who came later could read the story of their journey.

Sadly, thousands of pots and potsherds are removed from the land and sold for profit every year. Virtually all their potential information is destroyed when they are wrenched from their original locations, or proveniences. A pot in an art gallery or on a collector's mantle is all but useless if its provenience is unknown. Potsherds made into earrings or refrigerator magnets and sold in flea markets and online auction sites no longer have the ability to communicate about the people who made and used them.

As you visit archaeological sites and walk the ancient landscapes of the Hisat'sinom, please leave potsherds and other artifacts where you find them. Enjoy their beauty by picking them up and looking at them, but always place the artifacts back on the spot where they were resting. Do not place artifacts in a pile to help other visitors see them; this, too, disturbs the provenience, and it may lead others to help themselves to a souvenir. Taking or disturbing artifacts also violates federal, state, and tribal laws, and perpetrators are subject to fines and other penalties, including imprisonment. It is important not to purchase ancient pottery, even if it comes from private land. Doing so encourages additional looting.

Footprints of the ancestors were meant to teach us across the generations. If we recognize and respect that simple principle, they can continue to do so for a long time to come.

---

marks made by fire on their own local pottery. Many Hopi potters today say that the fire clouds and "blush" colors of plain pottery are even more beautiful than painted designs, because the marks left by the fire show that the pot is alive. Maybe the transactions that brought the decorated pots to Sierra Sin Agua villages helped maintain social and economic ties with distant neighbors. Pottery is just one of many items for which people of the region traded, and it is difficult to know whether the pottery itself or something contained in the vessels was the primary impetus for trade.

Neither Alameda Brown Ware nor San Francisco Mountain Gray Ware survived into historic times. But what happened to the people who made them? When archaeologists use ceramics as the basis for defining social groups, they can lose track of the makers when the ceramics no longer appear at archaeological sites. If some thirteenth-century Sinagua people moved to the Hopi Mesas, for example, did the potters among them abandon their distinctive pottery-making techniques and adopt Kayenta methods that were perhaps better suited to local sedimentary clays? Is the contem-porary Hopi aesthetic appreciation of fire clouds and orange mottling on their pottery a distant echo of Sinagua tastes for fire-clouded pottery, perhaps the legacy of long-ago immigrants to the Hopi Mesas? These questions deserve future attention, and collaborative research among scientists and Hopi artists might be a productive way to begin.

**Kelley Hays-Gilpin**, a professor of anthropology at Northern Arizona University and curator of anthropology at the Museum of Northern Arizona, has studied rock art and pottery in the Southwest for nearly thirty years. Among her many published books and articles, her *Ambiguous Images: Gender and Rock Art* won the 1995 Society for American Archaeology book award.

**Christian E. Downum** is a professor of anthropology in the Department of Anthropology at Northern Arizona University and former director of the NAU Anthropology Laboratories. He has conducted archaeological research in the Sierra Sin Agua since 1982, mostly at US national parks and monuments. He also serves as archaeological advisor to the Footprints of the Ancestors project, an intergenerational learning program that teaches Native American youths about the ancient places of the American Southwest.

Figure 15.1. Realistic depiction of a Gambel's quail on a Black-on-white Mimbres Style III bowl from Swartz Ruin.

# Expressions in Black and White

## Michelle Hegmon

Anyone who knows anything about Mimbres pottery knows that it is decorated with realistic paintings of animals and sometimes humans. Archaeologists, aided by biologists, are often able to identify what was painted on a pot—such as the Gambel's quail, *Callipepla gambelii*, complete with plume and dark throat. But identifying what is depicted is just one part of delving into the meanings of the pottery and its designs. Even contemporary artists often cannot say what a piece of art means, because it includes so many layers of nonverbal meanings. Instead of pretending to know what was in the minds of the prehistoric artists, archaeologists try to get at some of the possible meanings (the plural is important) of this special pottery and its significance to the people who made and used it.

The study and appreciation of Mimbres archaeology is a two-edged sword. Art historian J. J. Brody offered a comprehensive analysis of Mimbres pottery in his 1977 book *Mimbres Painted Pottery*, and I draw on some of his insights in this chapter. Brody found, however, that because of his book, many of the pots he depicted increased in value on the art market, providing an incentive for more looting and destruction of Mimbres sites. In his subsequent work, including a second edition of *Mimbres Painted Pottery* in 2004, Brody published only pictures of pots that were in publicly owned collections or museums and not on the market. We follow the same standards here.

### Origins and Use of Black-on-White Designs

Mimbres is part of a larger tradition that archaeologists call Mogollon, from the name of a Spanish governor. The label applies to a mountainous area spanning central Arizona (south of a massive geographical feature known as the Mogollon Rim), southern New Mexico, and northern Chihuahua. Mogollon pottery is typically red and brown. Because Mimbres pottery is so obviously black-on-white, it was once thought that the pottery was not Mogollon but derived from the black-on-white styles known in the northern Southwest. Mimbres pottery, however, was actually made of brown clay and was sometimes left brown on the outside. The pottery appears to be black on white because the makers covered the interior surface of bowls and the exterior of some jars with a thin layer (called a slip) of white clay, on which they painted black designs.

The black-on-white pottery tradition clearly had local antecedents. It began with a style known as Mogollon Red-on-brown, and then, in what archaeologists call Three Circle Red-on-white, the brown was covered by a white slip and the designs became finer. Eventually, the way people fired the pots changed so that the red paint turned black, and the Mimbres Black-on-white tradition was born. It started around 750 CE with a bold style now known as Style I. The most elaborately painted pottery is known as Style III, or Mimbres Classic, and dates from about 1000 to 1150.

Figure 15.2. An early Mimbres design reminiscent of decorations on Hohokam pottery from southern Arizona. *Left:* an early Style II bowl depicting a lizard; *right:* a similar lizard on a Hohokam bowl.

Among the most fascinating paintings on Mimbres pottery are representational designs—those depicting recognizable figures or scenes—although geometric designs were in fact more common. Representations of animals and other creatures began to appear, infrequently, soon after people started making black-on-white pottery in the 700s. They are similar to Hohokam designs from southern Arizona, with the animals incorporated into the overall design. Over time, however, the Mimbres style developed in unique ways. Sometimes the painter arranged two or more figures so that as the bowl was rotated, one replaced the other, in what is called rotational symmetry. Many of the bowls illustrated in this chapter display rotational symmetry. Most often, Mimbres bowls simply have a figure (such as a quail) or a scene as the central feature of the pot.

Mimbres styles became increasingly different from Hohokam styles at the same time the Hohokam tradition was expanding, around 1000 CE, suggesting that people in the Mimbres region were deliberately distinguishing themselves from the Hohokam and developing a more inward focus. It is this focus that we recognize as the Classic Mimbres phenomenon, and during its time, the Mimbres style was unique in the pre-Hispanic Southwest. Although other design traditions occasionally depicted animals, only Mimbres designs have creatures or sometimes humans as their single, central feature. At one level, then, the designs might have said, "We are the Such-and-Such people [we don't know what they called themselves], we live in this area, and we are different from people who live elsewhere and make different kinds of pots."

Mimbres painted pottery—mostly bowls, occasionally jars—is found in all sorts of contexts: on the floors of rooms, in trash heaps, and often in burials, placed over the deceased's face. In many cases, these mortuary bowls each have a small hole deliberately pecked out of the bottom, and often the small piece, akin to a doughnut hole, is also in the burial. These are sometimes called "kill holes," and it may be that they symbolically ended the life of the pot or allowed the spirit of the buried person to leave. The bowls are found with all kinds of burials—children and adults, men and women—and in almost all cases, there is just one bowl. It may be

Figure 15.3. Style III bowls with fish designs. *Left:* the long, toothed snout suggests that this fish is a gar, found in rivers in southwestern New Mexico. *Right:* the blunt head and corner mouth suggest a species of grunt found in the Gulf of California.

that the bowl people used most in life accompanied them in death.

Many of the bowls, including those accompanying the deceased, have wear on the inside and outside, indicating that at one time they were used for eating. Eating bowls would have broken occasionally, so each person probably used a number of bowls over the course of his or her life. At another level, then, the designs might have said, "I am me, the daughter of my mother, and this kind of design is appropriate for a woman of my age and standing." A new database, which I will describe shortly, will allow researchers to investigate this possibility systematically, by determining whether certain kinds of designs were associated with certain kinds of persons—women or men, for example, or adults or children. Preliminary research by archaeologist Stephanie Kulow has established that fish designs on bowls found with burials are consistently associated with adults and children but not with infants. She suggests that this is because fish have a special spiritual significance in contemporary Pueblo worldviews and infants' spirits might not have been ready to take on this significance.

**Studying the Designs**

Archaeologists can study the meanings of ancient pottery designs in at least two ways. The first is to consider what was painted and how it fitted into the painters' lives and beliefs. Were the creatures depicted ones they would commonly have seen or eaten? Did they copy styles from people in other regions? In the Mimbres case, do the images have any special significance in contemporary Pueblo religions?

The second way is to consider distribution—the counts of different kinds of designs across time and space. What is most common? Do certain areas or sites or kinds of burials have pottery with certain kinds of designs? Answers to these questions are knowable but, at this point in Mimbres research, mostly still unknown. This is because tools for these sorts of analyses are being developed.

In the 1970s, Steven LeBlanc organized the Mimbres Foundation to study what remained of Mimbres archaeology. As part of this work, he photographed many pieces of Mimbres pottery and recorded, mostly on index cards, what was known about where each vessel had been found. The photographs and cards became known as the Mimbres

Figure 15.4. Figures on Style III bowls combining elements of different animals. *Left:* birds with fish tails; *right:* four-legged creatures with human heads.

Archive. But because the data about where the pottery was found were not computerized, most researchers looked at what was painted in general and not at the spatial distribution of designs. For example, we know little about whether images of rabbits are found in one area or with one category of persons, and birds in another. In the past few years, LeBlanc and I, with the help of many other people, have added significantly to the collection of photographs and have put the photographs and data about the pots into a searchable electronic database known as the Mimbres Pottery Image Digital Database (MimPIDD), which is available on the Internet. With this new tool, it should be much easier to study the distribution of designs and thus investigate new levels of meanings.

In thinking about what the designs might have meant, it is important to remember a perspective that J. J. Brody emphasized: Mimbres paintings were the products of artists' imaginations, and they depict what the artists chose to represent. Even when the designs are realistic, we should not assume that they replicate the world the artists saw in their everyday lives.

A good example is the relationship between what the painters depicted and what they ate. We know that in Mimbres Classic times, most agricultural work involved growing corn, an important staple. But almost no images of corn (and few of any other plants) appear on Mimbres pottery. It is unlikely that the absence of corn in art means that corn was somehow unimportant to the artists; corn is a key focus of ritual for contemporary Pueblo peoples, and depictions of corn, sometimes realistic and sometimes stylized, are part of contemporary Pueblo art. We cannot say why the people of the Mimbres region ate but did not paint corn. It might be that corn held some special significance that somehow precluded its visual representation.

### Animal Depictions

Most Mimbres representational designs are animals, and they include a fascinating combination of realism, abstraction, and fantasy. Though all Mimbres wildlife images are stylized to some degree and most include some geometric fill, some of them (the Gambel's quail) accurately portray naturalistic details, almost like sketches in a field guide. Others (the turtles), are so stylized that they are only generally recognizable. In yet other cases, the depictions are of biologically impossible combinations, such as birds with fish tails, humans

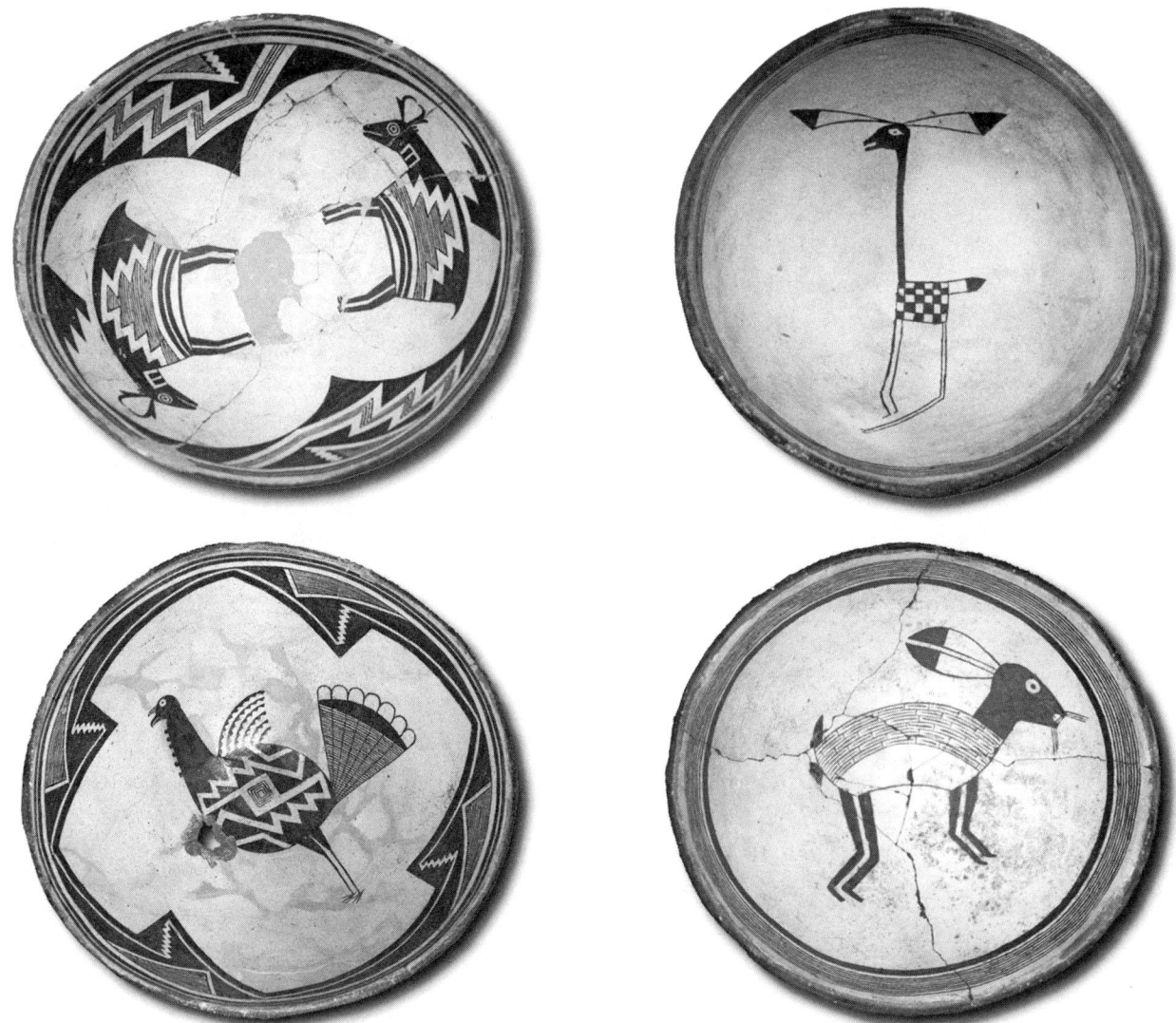

Figure 15.5. Style III bowls showing animals that were commonly depicted and often eaten. *Clockwise, from top left*: pronghorns (indicated by the antlers); stylized figure suggestive of a mule deer; jackrabbit (indicated by the black ear tips and long legs); turkey.

with four-legged bodies, and a rabbit with a rattlesnake tail.

In many studies of rock art around the world, archaeologists have concluded that although some of the wildlife depicted was also eaten, the art was not primarily about food or hunting magic. Art is more often about the way people understand the world and their place in it. The same is probably true of Mimbres pottery paintings. Some of the most commonly depicted creatures, such as quail, pronghorn, deer, rabbits, and turkeys, were also commonly eaten. We know this because their bones are found in trash middens. Others that were often depicted, however, such as swallows, bats, reptiles, insects, and fish, were probably seen often but rarely eaten. And animals that formed part of the daily diet, such as gophers and squirrels, were rarely depicted. It seems that the paintings did not mean "This is food."

What might they have meant? Studies of their distribution would go a long way toward answering this question. If certain kinds of wildlife depictions are grouped together at one site or area, it might mean that they were symbols of some kind of social group. We know that some of the depicted animals are Pueblo clan symbols today. Did they mean "We are members of the Rabbit clan"? To the contrary, we know of cases in which a burial was accompanied by several pots depicting different creatures, suggesting that the species were not simply clan symbols. Perhaps in a few years, with systematic study of the wildlife distribution, researchers will have more suggestions and better answers to offer.

As we move toward studying the distribution of different kinds of animal paintings, the way we classify the creatures will be critical. Many Western-trained archaeologists begin with the Linnaean system, which distinguishes scientifically named classes—mammals, birds, insects, and so forth—down to the level of species and subspecies. Certainly, Mimbres artists recognized some such distinctions, such as between creatures that lived on the ground and had fur and those that flew and had feathers, but other ways of classifying might have been more important to them.

Stephanie Kulow studied the distribution of wildlife imagery at the large Galaz Ruin. When she classified the creatures according to the Linnaean system, she found no patterns, but when she classified them according to their general habitat—earth, air, or water—she found that they clustered in different parts of the site. If the people who lived at Galaz grouped their symbols according to the creatures' habitats (or some other non-Linnaean criterion), then we will recognize the groupings only if we open our analyses to these other classification systems. Perhaps the paintings Kulow examined meant something like "We are the Water people, and they are the Earth people." Paintings that combine parts of different kinds of animals—different, at least, according to the Linnaean system—are fairly common on Mimbres bowls. It is important to remember that people in the past saw and classified the world differently than we do.

Figure 15.6. Style III bowls showing animals that were commonly depicted but probably rarely eaten. *Top:* swallow; *middle:* insects or possibly hummingbirds (the three-legged images are ambiguous); *bottom:* probably a caterpillar in the center of a flower.

The Mimbres region is biologically diverse; the nearby Bosque del Apache Wildlife Refuge, along the Rio Grande flyway, is a prime birding location today. Almost all the creatures depicted on Mimbres pottery are or once were found in the region, with two interesting exceptions. Peter Moyle, who has expertise in wildlife biology, identified many of the fish painted on Mimbres bowls as saltwater species, and at least one painting depicts a whale, complete with blow hole. The closest source of saltwater fish was the Gulf of California, some three hundred miles to the west across the Hohokam area. Assuming that Moyle's study is correct (it has been disputed), the artists either traveled to the coast or had close communication with people who did. This finding has implications for the gender of the pottery painters. In most societies in which pottery is produced on a fairly small scale, potters are women, although one person might have formed a pot and another painted it. It is certainly possible that women traveled from the Mimbres region to the Gulf of California, but ethnographic research suggests that such long-distance travel was more likely undertaken by men. Many scenarios are possible. Perhaps a few women pottery painters did make the journey. Perhaps male travelers gave women painters detailed descriptions of ocean fish. Or perhaps men painted at least some of the designs.

The second exception is macaws, the birds native to Mesoamerica that were brought into the Mimbres region. Formal burials of macaws have been found at some of the largest Mimbres sites, including Old Town and Galaz Ruin, and the red feathers of scarlet macaws are important in Pueblo rituals today. Many of the macaws painted on Mimbres bowls are shown being held and possibly trained by humans, indicating their special importance. When other birds or animals are depicted with humans, they are much more often shown as prey.

## Cosmology

Transformation, the continual process of being and becoming, is key to the way Pueblo and other Native American peoples view the world. A well-known example is the transformation of dead ancestors into *katsinas*, who return to the living as rain clouds. J. J. Brody, working with Rina Swentzell, from Santa Clara Pueblo, suggests that transformation is also a key theme in Mimbres painting and what we see as "combinations" (birds with fish tails, four-legged mammals with human feet) or even distortions of nature are actually transformations. Some bowls display pairs of combination creatures circling the pot. If this kind of rotational symmetry is found to be consistently associated with combinations, it would fit well with the idea of transformation. Such designs might have said, "In our world, being is always a process of becoming."

Some Mimbres paintings depict deities and other figures important in Pueblo and Mesoamerican cosmology, including serpents, emergence scenes, and masked figures. There are also death-related paintings, including decapitation scenes, animals of the night, such as bats and owls, and the Hero Twins, who conquered the lords of the underworld in Maya mythology. Even apparently ordinary animals may have cosmological significance. Rabbits, for example, are linked to the moon, and many rabbits have lunate bodies. It seems likely that at a very general level, the paintings meant something like "With this pottery, we express our understanding of the world" or "With this pottery, we express reverence for our deities." Once again, a better understanding of the distribution of these kinds of images will shed a great deal of light on their significance in Mimbres culture. For example, it might be that the practice of using masks spread from Mesoamerica to the Pueblo world through the Mimbres region. Did paintings of masks first appear in the southern Mimbres area or in the travel corridors? Are the images of deities found in what appear to have been special places? Did only some people have access to them? Fortunately, future research is likely to answer some of these questions.

## Humans

Perhaps no Mimbres designs have drawn as much attention as those depicting humans (plate 12). Most of the humans are shown as parts of active scenes, telling us something about past lifeways—or at least those the artists chose to represent. Sadly,

Figure 15.7. Bowl designs showing animals of the night, which might have had cosmological significance as symbols of death or the underworld. *Left:* a bat wearing a mask across its eyes; *right:* an owl, seen from the perspective of its prey below.

many faked paintings of humans exist, and because they are often painted on real Mimbres pots, they are almost impossible to detect analytically. As a result, systematic studies of Mimbres paintings of humans are difficult. When done carefully, however, they can provide many insights.

One of the most controversial studies of Mimbres pottery is an article I wrote with Wenda Trevathan, a biological anthropologist who studies the evolution of human birth. Trevathan had lectured in my class on gender in archaeology, and she told the students how human infants, in contrast to infants of most other primates, are normally born facing backward, meaning that the mother needs help from others to guide the baby out. Several weeks later, I showed slides of a Mimbres birth scene in which the infant is facing forward. The students immediately pointed out that the birth was "wrong." Trevathan and I found several other, similar Mimbres birth scenes and wrote an article in which we suggested that the scenes had been painted by men, because men were less likely to be well informed about the details of birth.

The article was published by a major journal, and the response was immediate and furious: We were ignoring the possibility of artistic license; we were ignoring cracks on the bowls; we were ignoring other birth scenes (which we had not included because we thought that they were fakes); we just *had* to be wrong. I do not know for sure whether we were right or wrong, but now, with plenty of hindsight, I laugh over the response. I think that what raised so much ire was not the conclusion itself—it is perfectly sensible to suggest that men might have done at least some of the painting—but our reasoning. What made people so angry was that we based our conclusion on the assumption (which I still think is reasonable) that men were uninformed about birth. Interpreting art is difficult.

**Understanding Mimbres Art**

One of my favorite ways of understanding art is based on a quip by the modern dancer Isadora Duncan: "If I could tell you what it means, there would be no point in dancing it." Art cannot be reduced to words, and frustrating as that might seem for writers, it is probably a good thing. More research with a systematic database will enable

Figure 15.8. Depiction of a birth scene on a Style III bowl from Swartz Ruin.

archaeologists to evaluate some of the possibilities I have suggested in this chapter, such as the possible relationships between designs and individual or group identity. Doing so will go far toward enhancing people's understanding and appreciation of the pottery. But even as we hone in on specific meanings, we must keep the plural in mind. The designs probably had many meanings, ranging from aesthetics to identity and cosmology, and probably the intersection of those meanings forms yet another level of meaning. Perhaps the greatest appreciation of Mimbres art will be found in this complexity.

**Michelle Hegmon**, a professor of anthropology at Arizona State University, is best known for her work on pottery and the archaeology of the social realm. She is the editor of *The Archaeology of Regional Interaction: Religion, Warfare, and Exchange across the American Southwest and Beyond* and co-editor with Margaret C. Nelson of *Mimbres Lives and Landscapes* (SAR Press 2010).

Figure 16.1. Ghost-like and other Basketmaker anthropomorphs, Canyon de Chelly, Arizona. Two smaller white birds (possibly quail) are painted over a red human form.

# Ancestral Pueblo Rock Art in Tsegi Canyon and Canyon de Chelly
## A View behind the Image

*Polly Schaafsma*

> *"Early Basketmaker life and customs, as revealed by archeology and interpreted in the light of modern Indian life, [make] an unsatisfactory picture at best. That is the weakness of archeology. It tells much of the life of the body and the work of the hands, little of the life of the spirit and the work of the mind."*
> —Charles Avery Amsden, *Prehistoric Southwesterners from Basketmaker to Pueblo*

Charles Amsden's statements, part of a detailed and sensitive synthesis of Basketmaker life published in 1949, express his frustration with the lack of archaeological evidence concerning the spiritual life and ideologies of prehistoric peoples. Amsden likens the archaeological record to "a painting [that] can give us no more than the eye can see and the imagination conjure from its revelations." He senses that "the Basketmakers [and the ancestral Pueblos in general] were as active in mind as they obviously were in body."

Nevertheless, certain bits of evidence in material culture point to various aspects of ancestral Pueblo religious practices. This evidence includes rock art. Various types of studies have shown that we can find in the imagery of rock art indications of the ideological dimension of ancestral Pueblo life. Granted, when we marshal all of the evidence available, we may glean only a glimpse of this spiritual and mental realm; however, such a glimmer of insight is much better than nothing at all.

Rock art adds a tantalizing dimension to the canyons and alcoves of the Colorado Plateau in northeastern Arizona. Tsegi Canyon and Canyon de Chelly are located in the central San Juan region of the plateau, an elevated tableland dissected by deep, colorful sandstone drainages. *Tsegi*, a Navajo word meaning "within the rock," and *Chelly*, a corruption of this same word, are appropriate names for these canyons, with their vertical walls offering protected arching alcoves and rock shelters, which provide a dramatic setting for the rock art. Tsegi Canyon on Laguna Creek, a tributary of the Chinle Wash, and Canyon de Chelly on the Chinle itself, share common rock-art styles representing 1,300 years of ancestral Pueblo occupation. The same styles are also found on the San Juan River itself and in nearby canyons entering the San Juan from the north, such as Grand Gulch, Slickhorn Gulch, and Butler Wash. Through the centuries, this art underwent gradual and perceivable changes in style, symbolism, and meaning that paralleled changes in cultural needs and values. The styles can be dated relative to one another by various instances of superimpositions and also on an absolute basis by their associations with habitation sites.

Various estimates place the earliest Basketmaker occupation at about 1 CE. It lasted until around 450 CE. The rock art of this period is characterized by pecked and/or painted representations of large anthropomorphic figures with squarish or broad-shouldered, tapering bodies and drooping hands

Figure 16.2. Large Basketmaker II figures superimposed by birds and a smaller, bird-footed anthropomorph in red. A line of red ducks appears at the upper right. Blue Bull Cave, Canyon del Muerto.

and feet. These figures may wear necklaces and sashes, and their torsos may be dotted or divided by vertical or horizontal zigzags. Headgear, when present, is conspicuous and varied. At painted sites, handprints are often numerous around the anthropomorphs. At late Basketmaker II sites in the northern San Juan, birds are sometimes located on the heads of anthropomorphs. Zigzags or snakes, atlatls and/or spears, and medicine bags or pouches are other associated elements. Archaeologists Alfred Vincent Kidder and Samuel Guernsey, working in the Tsegi region early in the 1900s, were the first to observe that the distinctive, broad-shouldered anthropomorphic forms in the rock art were consistently found in Basketmaker II shelters.

Basketmaker rock art is related, at least conceptually, to earlier Colorado Plateau rock-art styles found immediately to the north. These earlier styles emphasize large, abstract anthropomorphs

Figure 16.3. Basketmaker woman in white and probably a shaman carrying a wicket-shaped object. This scene seems to be related to another in the same shelter, in which a reclining female is administered to by someone holding over her a similar device. Canyon de Chelly.

with.,supernatural attributes. The figures may represent shamans, supernaturals, or supernaturals seen by shamans. In rock art of the Barrier Canyon

Figure 16.4. Positive striped prints and a single negative handprint on a rock shelter ceiling in Canyon de Chelly. Ancestral Pueblo period unknown.

Figure 16.5. Early ancestral Pueblo paintings of avian and lively human forms, a line of triangles, dot patterns, and double wavy lines in red and white. A white bird-headed man appears in the lower right. The panel is in Canyon del Muerto.

style, quadrupeds, birds, and snakes occur around these figures in a manner suggesting that they are spirit helpers or guides.

The early styles shed light on the significance of the Basketmaker rock art, which they influenced. Like the earlier anthropomorphic figures, the

Basketmaker figures are not ordinary men. They are abstract, remote, static forms lacking human qualities. Their elaborate headgear is another clue to their supernatural affiliations. The headdresses may consist of tall terraces of lunate elements or tall feathers. Sometimes these elements are shown projecting from the left ear, as well as from the top of the head. Occasionally, staring eyes and unnatural hands and feet are pictured. The latter may be three toed and bird like or resemble clawed bear paws, both of which suggest nonhuman affiliations. Numerous handprints, birds, and occasional snakes are closely associated with these ethereal beings. The handprints are usually stamped near the figures, or they may be placed on their torsos or around the heads. Occasionally, babies' footprints were also stamped on the rocks. The consistent association of handprints with the anthropomorphs at these Basketmaker sites suggests that they were made for some definite purpose. They may have been left as signatures of prayer requests or made in the act of obtaining power—either from the rock-art figure itself or from the place it occupied. In both cases, some kind of supernatural power associated with the human forms is implicit.

Both birds and snakes are almost universally associated with shamanistic iconography and are viewed as possessors of certain types of wisdom and specific powers. Both function as messengers to other realms. Snakes have access to the underworld and symbolize regeneration and rebirth, while birds, on the other hand, are symbols for freedom and spiritual transcendence. Magical flight in bird form is common to shamanistic traditions not only in the Southwest but also worldwide.

At some point, probably late in the Basketmaker II period, birds became very important in the iconography of the canyon rock art. These birds are most commonly depicted as rather stylized, round-bodied forms without wings but often with legs drawn at an angle to convey a sense of flight. As paintings, they are often done in white with red outlines, or vice versa. The heads and necks were sometimes painted in fugitive pigment, so today these birds appear headless, although this was not originally the case. Present in great numbers from late Basketmaker II into Basketmaker III times, these birds

Figure 16.6. Ancestral Pueblo stick figures with large birds on their heads, Kiet Siel, Navajo National Monument.

Figure 16.7. Ancestral Pueblo petroglyph of anthropomorph with bird symbolism, Marsh Pass.

are often readily identifiable as ducks, turkeys, and occasionally geese.

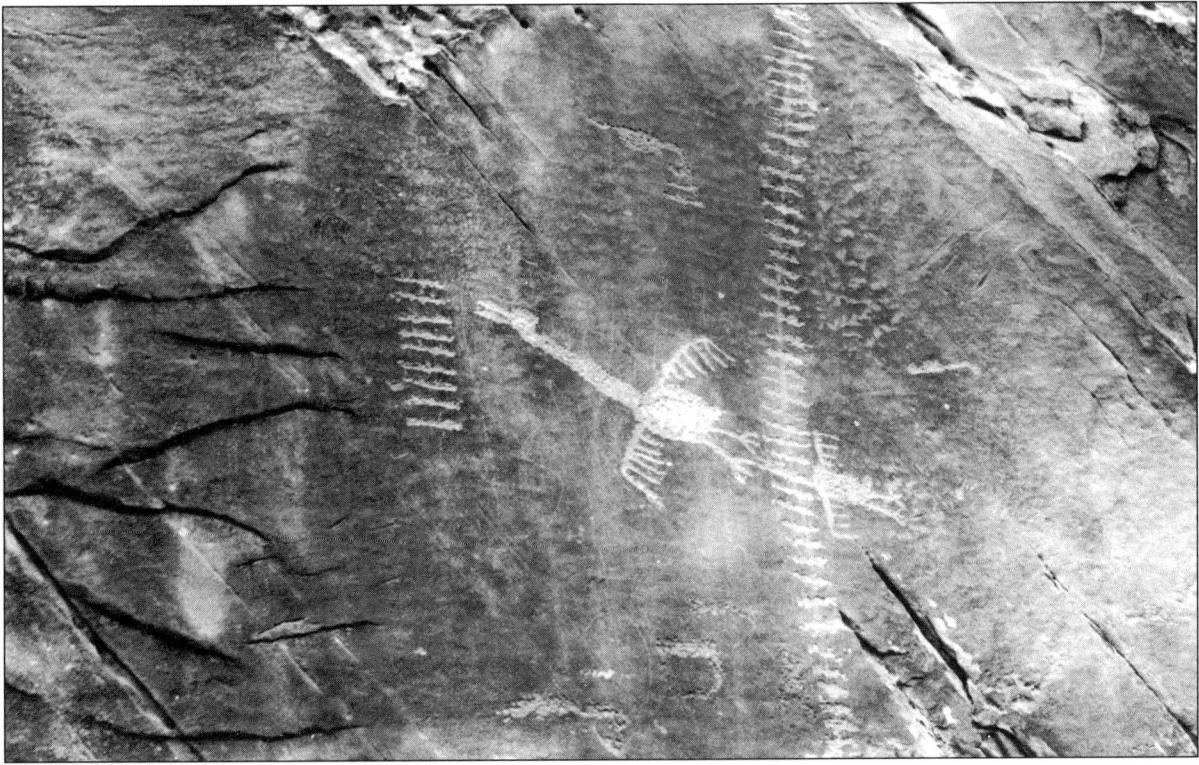

Figure 16.8. Petroglyphs of ducks or geese in flight, Marsh Pass below Tsegi Canyon. The significance of the sequences of short lines is not known.

Although the symbolic import of these different species of birds is difficult to determine for the Basketmakers, their significance for the modern Pueblos is instructive. Birds in general are an extremely important element in modern Pueblo ritual and mythology, and judging from their appearance, not only in rock art but also on ceramics, in kiva murals, and as fetishes, they enjoyed this status continuously throughout hundreds of years of prehistory. Mobile creatures that travel swiftly, birds frequently play the role of messengers in Pueblo myths. Feathers that can float on air are analogous to thoughts, and they are associated with prayers that must travel to the spirit world, to the powers controlling the rain and clouds. Thus, feathers are commonly used in constructing ritual paraphernalia such as prayer sticks.

The duck's ability to be at home everywhere, from high in the air to under the water, gives him extraordinary powers. In both Mexico and the Southwest, he is ascribed shamanistic attributes. Among the contemporary Pueblo Indians, he is thought of as a ventriloquist. He is regarded as wise because he is a great traveler and searcher, migrating with the geese and cranes, to which are assigned similar powers. The duck is viewed as a keeper of myths, and like other birds, he functions as a messenger. Supernaturals are said to assume duck form in their travels between this world and the spirit world. The water associations of the duck are obvious, and he can carry messages to the clouds, the sources of rain. These characteristics and abilities are not inconsistent with the representations of the duck in ancient rock art, where he occurs near or on, or replaces the heads of, figures that may represent shamans in spirit flight.

The turkey, a creature of the earth (as opposed to the sky), is bound to embody a different set of symbolic concepts, although this bird, too, is found in the same set of relationships to human figures in the rock art. The turkey was domesticated by the ancestral Pueblos by 700 CE. Its feathers were used for robes and ritual purposes at least by that time and perhaps even earlier. Among the modern Pueblo Indians, the turkey is symbolically associated with the earth, springs, streams, and mountains,

Figure 16.9. Ancestral Pueblo human figures in red and white, Canyon de Chelly.

which are the homes of the cloud spirits. It follows that the turkey is viewed as an intermediary between these mountain water sources and the rain clouds that form on the peaks. He is also regarded as a teacher and helper, and he is associated with the dead, who must return to earth before rising as clouds to the spiritual realm. Turkey feathers are therefore used in mortuary offerings. Prehistorically, turkey burials have been found with human burials, and the practice of wrapping the dead in turkey feather robes may have been a means of assisting the dead in their spiritual journey. No research has been done to see if the turkey-headed anthropomorphic forms in rock art occur in caves containing burial cists; if such a connection were found, the rock art would also depict this relationship.

This excursion into Pueblo ideology provides a direction for understanding the ancient paintings and petroglyphs of Tsegi Canyon and Canyon de Chelly. The importance of bird symbolism in Basketmaker ceremony and art is further emphasized by some of the artifacts retrieved from Basketmaker sites. A stuffed bird skin and a bird-headed wand with feathers and bird tails attached were found in White Dog Cave, a Basketmaker II site near Tsegi Canyon. Skin medicine pouches containing feathers have also been found at Basketmaker sites.

By the time birds became abundant in the rock art of Tsegi Canyon and Canyon de Chelly, other major changes were taking place. The strict canons of Basketmaker II painting were relaxed. The austere figures of the earlier period were gradually replaced during an efflorescence of rock art in which a variety of smaller images were painted in bi- or polychrome designs. This post–Basketmaker II period was truly the heyday of rock art in Canyon de Chelly. Individuality of expression and the portrayal of unusual figures were the name of the game. There was great experimentation with the human form, which continued to be a dominant image in quantity if not in size. It was rendered in a variety of shapes including small rectangular-, trapezoidal-, and triangular-bodied forms, stick figures in frontal poses, and stick figures in profile—any one of which might appear with a bird on its head. An occasional bird-headed figure is shown with legs bent, as if actually flying through the air.

Figure 16.10. An early ancestral Pueblo figure in red plays a white flute below a rainbow while two quadrupeds frame an ambiguous object at right, Canyon de Chelly.

Figure 16.11. Pueblo III painted textile design in white and reddish brown, Canyon de Chelly.

There are lines of hand-holding men and women, dancing groups, Siamese twins, and seated stick figures with headdresses or headbands, playing flutes under rainbows. Others hold huge arrows or ceremonial staffs. An occasional crane was added to the avian inventory. Bizarre insects, a few nondescript animals, and mountain sheep with open mouths are represented. According to Campbell Grant, in his book *Canyon de Chelly: Its People and Rock Art*, this complex dates between 450 and 1100 CE. It is less well represented in the Tsegi, although some paintings in the vicinities of Kiet Siel and Scaffold House seem to date from this period.

The flute player is prominent in the rock art of Tsegi Canyon and Canyon de Chelly beginning in Basketmaker times. According to Grant, the earliest flute players in Canyon de Chelly are simple stick figures. The seated flute players under rainbows

Ancestral Pueblo Rock Art 143

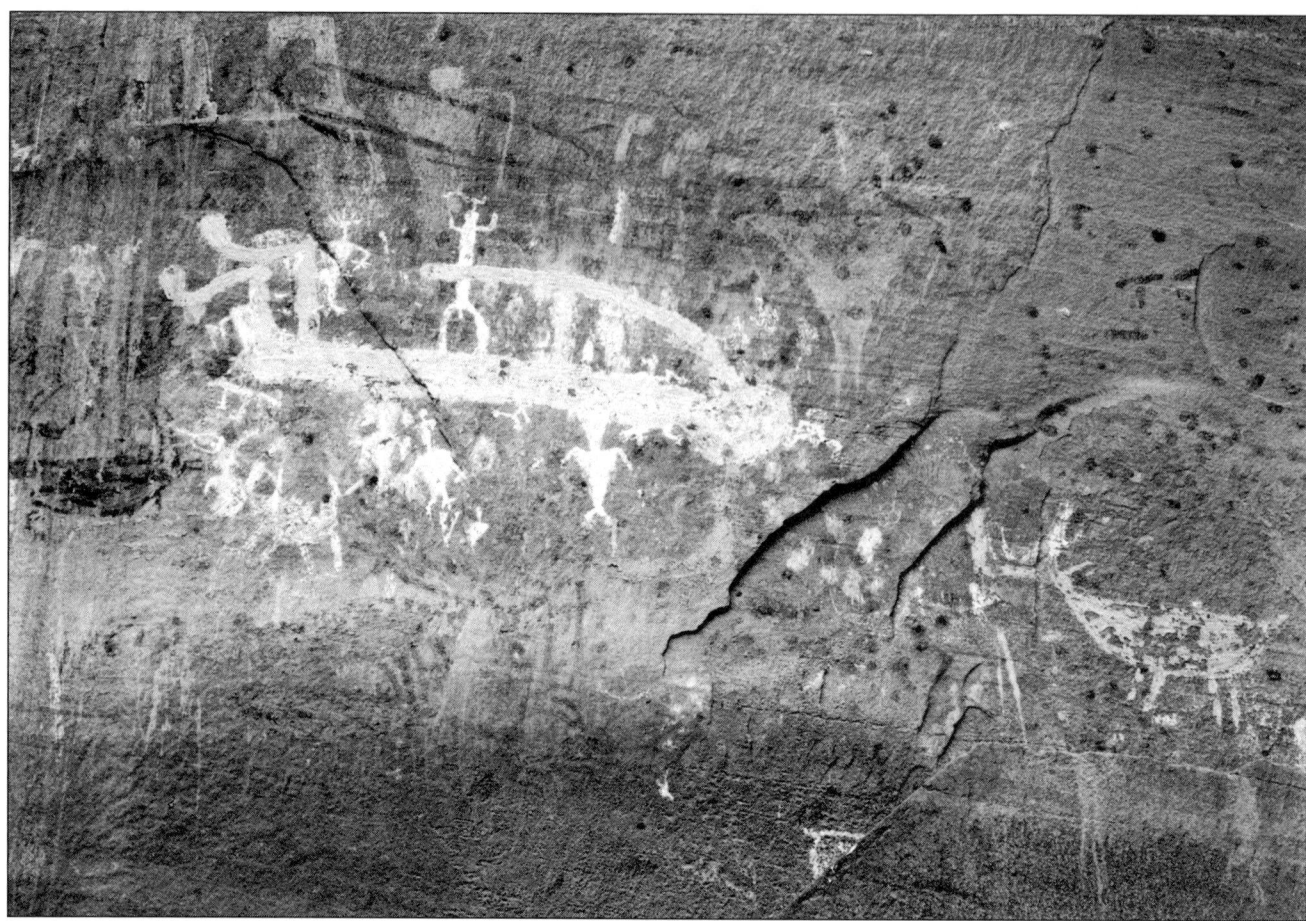

Figure 16.12. A five-foot-long flute player smeared in white clay dominates this painted portion of a shelter wall in the Tsegi Canyon system. The Tsegi Phase flutist is painted on top of a myriad of earlier ancestral Pueblo elements painted in pastel clays. Ancient Basketmaker men and two stirrup-shaped elements loom ghost-like behind the later figures.

may be early images of this personage as a rain priest. Later examples in Pueblo II and III rock art are more apt to be shown as phallic and/or humpbacked and even kicking their heels. In his more complex manifestations, the flute player may be associated with snakes or appear in hunting scenes with mountain sheep, where he seems to function as a hunting magician. In still other instances, he takes on insect attributes and is shown with an insect-like hump and antennae. A most unusual flute player is the five-foot-long painting in white clay in Flute Player Cave in Tsegi Canyon. This is a reclining figure without phallus or hump. He is superimposed on a wealth of previously painted designs and is attributed to the Tsegi Phase (1250–1300 CE), the latest Pueblo occupation in the canyon.

This distinctive and conceptually varied character, who has always been of interest to southwestern scholars and aficionados alike, appears on ceramics and in kiva paintings, as well as in rock art. Like all figures rooted in Pueblo myth and ceremony, his meaning is complex. To this day, he has persisted in the Pueblo pantheon. In his various manifestations, he plays several distinctive but interrelated roles. As a rain priest, he plays his flute over springs, and with the help of toads and insects, he attracts clouds and moisture. He also plays his flute while preparing magic locust medicine for purification rites. His association with mountain sheep gives him a special relationship to the Horn Clan at Hopi. An overriding characteristic of this fellow is his sexuality and his role in fertility rites. His hump is said to be filled with babies, as well as seeds, belts, and blankets—gifts for the maidens he seduces. Klaus Wellman, who studied the symbolism of this figure in some depth, has suggested that this little

Figure 16.13. Tsegi Phase mountain sheep, handprints, and a large circular painting in white. The latter is interpreted by the Hopi as a Fire Clan symbol. Betatakin, Navajo National Monument, Arizona.

Figure 16.14. Early ancestral Pueblo figure with supernatural qualities seemingly indicated by wavy lines that emerge from either side of his head. Four stick figures in red appear to float between the zigzags. A large shamanic form with necklace appears at the right. Canyon de Chelly.

personage is a regional variant of a more comprehensive archetype, the Universal Trickster. The Trickster, present in mythologies throughout the world, embodies unprincipled, amoral forces and chaos on one hand, while on the other, he becomes the creator, cultural hero, and transformer. These are universal roles that the later Pueblo Indians also dramatized as the Horned Water Serpent and clowns.

The final phase of ancestral Pueblo rock art in the central San Juan (Pueblo II to III, 1000–1300 CE) had its own distinctive features. Both petroglyphs

Figure 16.15. Stamped Pueblo handprints in red, bordered above and below by broad bands in white and yellow, respectively, Bubbling Spring Canyon, Tsegi drainage.

and large paintings in clay were popular. Around 1100 CE, representations of cotton textiles and textile designs are a significant component of the rock art in the Little Colorado River region and elsewhere, and a few are represented in Canyon de Chelly, such as the one pictured on page 143. These elements are believed to have had significance because of the ritual link between cotton and clouds. These patterns may have been made to attract rain clouds, hence being viewed as petitions or prayers for rain.

Flute players still appear, and mountain sheep and hunting scenes are also typical of this late phase of canyon rock art. The practice of making handprints with clay and mineral pigments persisted throughout the ancestral Pueblo occupation, and hand stencils are characteristic, as are lizards, lizard-men, and rectilinear stick-figure humans. The human figures tend to be angular and geometric.

Particularly notable are the large, white, circular designs painted in shelters with cliff dwellings. These prominently located paintings are not only highly visible but also bold in concept. They consist of a variety of simple patterns that sometimes utilize negative designs. Others are variations on the concentric-circle motif. The large circle painting at Betatakin contains a negative image of an anthropomorph, and arcs in red and yellow are painted at the base and on either side. The figure has been interpreted by the Hopi as a representation of Masauwu, god of the earth, guardian of the dead, and controller of fire, and as a Fire Clan symbol.

I previously made the suggestion that these large white paintings were emblems of the social groups who occupied the sites. Certainly, their large size and conspicuous locations would announce this information to outsiders. Jane Young, in her research with the Zuni, found that it is considered good luck to have one's clan symbol about, and this may have been equally true in the past. However, taking into consideration the defensive locations and broad social contexts in which these paintings are found, I now believe it far more likely that they represent a visual form of defensive magic during times of stress, serving as warnings or possibly even threats to marauding groups. Ethnographic information supports the idea that shields and their designs were perceived as having magical protective properties above their physical usefulness for defense.

Other rock art, both in the form of petroglyphs and small paintings within these alcoves, occurs on the cliff walls near rooftop work areas of family dwellings. They are usually casually made and seem incidental. One set of four white handprints in Inscription House in Navajo Canyon occurs directly above a wall that divides family living areas, and these prints may have served as a social boundary marker within the village.

Quadrupeds such as mountain sheep, which are present throughout all periods of ancestral Pueblo rock art, become more prevalent during Pueblo II and III. It is obvious from the earlier discussion that representations of animals are usually more than simple depictions of the natural fauna and that they were often selected for their symbolic significance and the particular powers attributed to them. Horned animals almost universally are viewed as "powerful," and the mountain sheep in the Southwest is no exception. The mountain sheep is important in myth, and his horns are worn as symbols of power and knowledge by Hopi priests of the Horn Society. Deer and mountain sheep are also frequently represented in hunting scenes. It is probable that this imagery was made in a ritual framework in which prayers and offerings were made either to ensure a successful hunt or to propitiate the spirits of animals already slain.

Lizards and the ambiguous forms of lizard-men are prominent themes in late ancestral Pueblo art. The lizard, like the bird and the snake, is a typically shamanistic motif, symbolizing bodily transcendence and rejuvenation. Exactly what the lizard signified to the canyon artists is not known, although the existence of a definite conceptual role in the ideology of the times is reinforced by the stone lizard-woman effigy excavated from a Pueblo III kiva at Salmon Ruin on the San Juan River near Bloomfield, New Mexico. According to Young, to contemporary Zunis, petroglyphs of these figures near Zuni Pueblo represent "raw beings," "the way the Zunis looked at the time of the beginning; 'before they were finished,'" when they had tails and webbed hands and feet.

Thirteen hundred years or so of changing ancestral Pueblo ideas are represented by these enigmatic graphic images on the red sandstone walls of Tsegi Canyon and Canyon de Chelly. I have touched on salient characteristics and shifts through time, pointing out a few of the possible and probable implications of the rock art. Concerns for rain are pictorially evident throughout, and one of the last concerns was that of defense, just prior to migrations away from the Four Corners region. The art also suggests that the early occupants of these spectacular red canyons were concerned with a world of spirit powers, to which they related through shamanistic specialists who had access to these realms. The "life of the spirit and the work of the mind" of the ancestral Pueblo are revealed by their graphic imagery as a complex world of symbol and metaphor. It reflects a way of relating to the environment that is somewhat unfamiliar to our "scientific" rules and biases, but one that was equally meaningful and all-encompassing.

**Polly Schaafsma** is an anthropologist and a research associate at the Museum of Indian Arts and Culture, Museum of New Mexico. She is the author of many books, including *Indian Rock Art of the Southwest* and *Images and Power: Rock Art and Ethics*.

Figure 17.1. Zig-Zag Mountain in the Santan Range, Gila River Indian Community.

# Songscapes and Calendar Sticks

*J. Andrew Darling and Barnaby V. Lewis*

When was the last time you broke into song? Which song did you sing? Here is an O'odham song about a mountain that stands west of an old Indian trail. Nameless on maps, it is a monument of volcanic and metamorphic rock that is well known in O'odham country:

> Zigzag Connected,
> On top I pause.
> Here beside me,
> Black cloud floats zigzags,
> Pleasant for watching.

Or so says a translation by the anthropologist Donald Bahr. It and translations of other songs in this chapter are drawn from the 1997 book *Ants and Orioles: Showing the Art of Pima Poetry*.

The late Vincent Joseph, an O'odham elder who recorded the song in the 1980s, offered his interpretation: "The Oriole bird, the traveler, while resting on top of Zig-Zag Mountain, sees a black cloud floating below. The cloud imitates the zigzag shape of the mountain, and the Oriole is pleased with what he sees."

Twenty years later, Barnaby Lewis shared the same song as we traveled along a historic O'odham trail on land belonging to the Gila River Indian Community. The morning was not unusual. Our jobs for the Tribe had brought us to inspect recent vandalism to rock art sites on the reservation's northern boundary in the Santan Mountains. From the west, Zig-Zag Mountain stood out clearly. Barnaby's singing blessed the moment and honored the spirit of the mountain, even through the windshield of a tribal vehicle.

This chapter is about the way in which descendants of the Huhugam (Hohokam) interpret geographical space through song traditions. We examine songscapes—landscapes remembered through O'odham song—and their relationship to traditional infrastructure for travel and the archaeology of ancient trails. We also consider the dimension of *time*. Time is important for understanding how landscapes, particularly, sacred landscapes, exist alongside history. If we wish to appreciate the traditional O'odham's spatial concepts, then we must consider them in the context of O'odham ideas of time and history, specifically the histories told through O'odham calendar sticks.

## Trails

Trails are a major part of traditional infrastructure. The arid Southwest offers a unique cultural landscape in which trail segments remain visible for a long time on the desert surface. Some are as old as ten thousand years. The products of regular foot travel, desert trails appear as scars in the natural desert pavement, unlike roads or engineered constructions. Native infrastructure includes both the facilities, such as trails, and the ideas that enable communities to function. Knowledge of trails in the past—not just of where people were, but how they got there—is important for archaeologists' understanding of the locations and distribution of sacred sites and settlements.

For the O'odham of central and southern Arizona, traveling means more than going from one place to another. Travel is not random. Traditionally, a person can travel on foot and through dreaming.

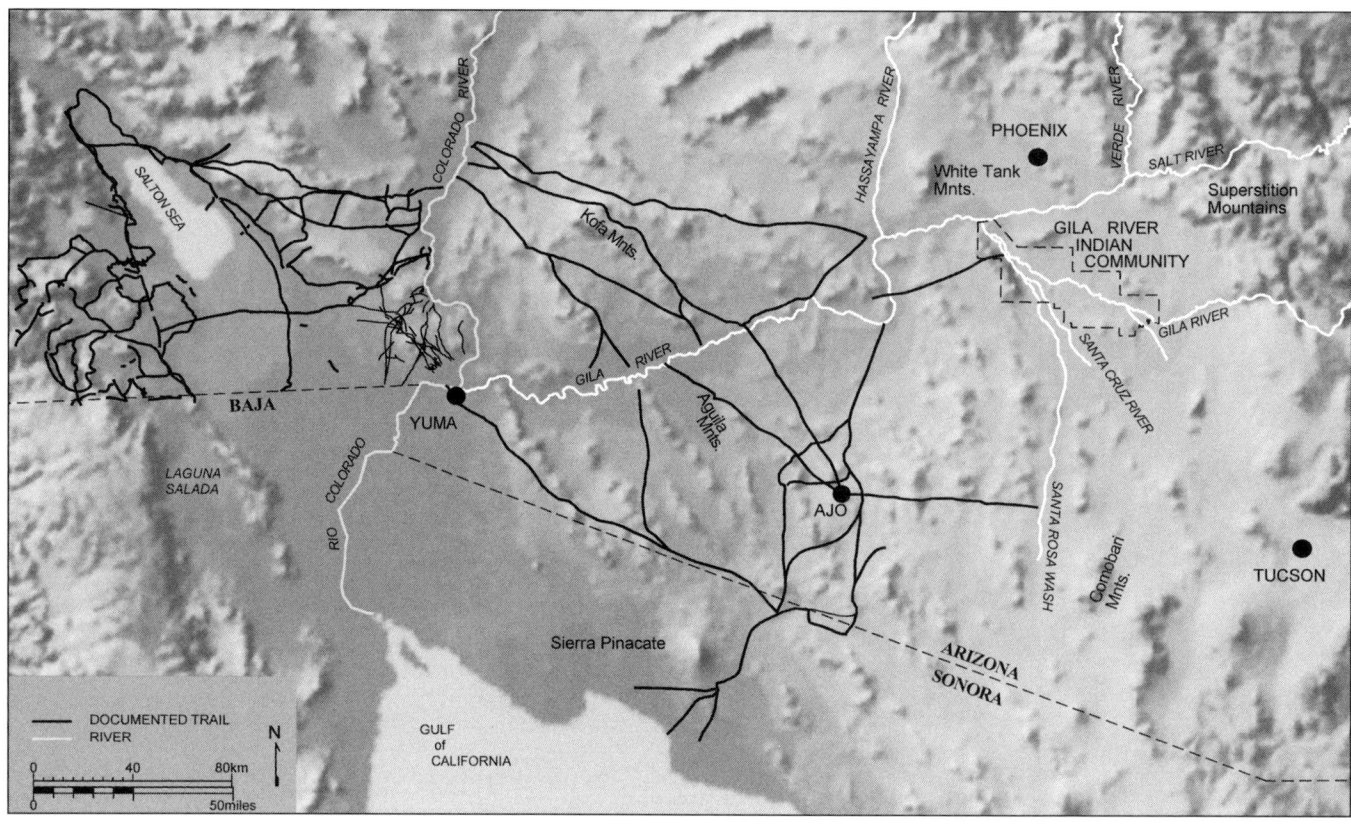

Figure 17.2. Major Native American trail networks in the southern US Southwest and adjacent northwestern Mexico.

An individual acquires songs and spiritual power through dream journeys to spiritual places. Such journeys recapitulate physical travel over the ground and underlie the relationship between beliefs about travel and the actual trails that certain songs or song series represent. In the physical world, journeys must be enacted with geographical precision and spiritual propriety, vital components in surviving desert travel. The same is true for singing songs: They must be sung in the correct order. The repeated performance of O'odham songs expresses and reinforces the experience of travel and the qualities of places in the O'odham world. In this way, the O'odham may know a place through their song culture without having traveled there. In a world without maps, a person who knows the songs is a person who knows where he or she is going. More important, a person who knows the songs knows the dangers along the way.

## O'odham Social Singing

O'odham social songs, such as the Oriole, Swallow, and Ant songs, accompany traditional dances held at all-night social events called sings. Whether sacred or secular, songs are never created by singers. Instead, the animal spirits, often birds, compose them and often teach the songs to a singer through dreams. Once a singer has learned a song from an animal spirit, he or she can pass it to other singers through performance and imitation. Songs are typically short and are arranged in an ordered series consisting of as many as a hundred individual songs.

Many of these songs have to do with places. One song from the Oriole series, for example, describes the Oriole bird's unsuccessful attempt to enter the mountain known as Crooked Red:

> Red Bent, Red Bent.
> Inside songs sound,
> And I am poor [sad],
> I circle behind, Oh, what can I do?
> Now, enter, and then, many songs know.

Often, a portion of a particular song series describes a journey, each song representing a place along the way. The geographical aspects of sings

Figure 17.3. Faces on Crooked Red Mountain, Gila River Indian Community. Mountain spirits reveal themselves to remind onlookers of their presence.

and their relationship to archaeological features, especially trails, interest us most here. Donald Bahr observed in *Ants and Orioles*: "It is well to think of social dance song sequences as postcards sent from someone on an impassioned journey. On receiving the card one speculates about the mood of the sender, about all that was happening at the moment of the message…and what the next step in the journey might be."

In a song series, geographical references to place follow a linear, circular, or meandering path without repeated returns to the same location. The sequence of songs creates the geographical itinerary of a sing. The Oriole songs serve as a particularly good example that can be tied to the archaeology of ancient trail systems.

## Oriole Song Archaeology

The Oriole song series as rendered by Vincent Joseph consists of forty-seven songs. Fourteen of them, songs nine through twenty-two, describe a journey to the salt flats on the Sonoran Gulf Coast, a circuitous route along which the traveler visits sacred mountains and hot springs. When archaeologists look for evidence that prehistoric people took a long journey similar to this, they find traces of ancient trails leading from the Gila River Indian Community to the Gulf of California, covering a distance of approximately 286 miles. These trails link the places named in the Oriole songs. Consequently, we can relate specific trails found during archaeological surveys to the itinerary described in the song series.

The westward journey begins by heading north, starting at Blackwater Lake in the Gila River Indian Community and continuing to the Santan Mountains on the reservation's northern boundary. There the songs describe several mountains: White Pinched, Zig-Zag, Crooked Red, and Long Gray. Archaeologists have successfully identified several trails in the mountains, and just as the song sequence directs, they head northeast toward the next destination, the Superstition Mountains. Archaeologists have not traced the trail past the Santans, but the Superstitions are easy to identify and offer the traveler a clear point of orientation.

What do these old trails look like? It depends on local conditions, but intact trails are normally twelve to twenty inches wide. Where they cross desert pavement, they become a clear track or multiple parallel tracks worn as much as two inches below the surface. As the trail rises into the Santan

Mountains, it is sometimes deeply incised and shows signs that someone purposefully removed and stacked rocks to clear the way. We also find artifacts, mostly fragments of pottery jars—clear evidence that the Huhugam ancestors of the O'odham used these trails centuries ago, between about 950 and 1450 CE.

After leaving the Superstition Mountains, the Oriole song itinerary proceeds westward through a series of mountain destinations: Iron Mountain, Thin Mountain, and finally, turning south, South Mountain. Archaeologists find scant evidence of a trail connecting these places until they reach the westernmost ridges of South Mountain. According to the Oriole songs, from there the trail crosses the Gila River to a place on the west side of the valley. At that point, visible trails head west through a pass in the Estrella Mountains. All along this part of the journey, we find rock art and trail shrines.

On the eastern side of the Estrella Mountains, across a high pass and for a distance of about twenty-two miles, we find archaeological evidence of a major trail to the west crossing the Rainbow Valley into a pass in the North Maricopa Mountains. It effectively crosscuts the large S-curve in the Gila River. The Spanish missionary Father Kino traveled this part of the trail once in the winter of 1699, but it would have been a difficult route for Europeans burdened by draft animals and equipment.

Field identification of portions of the trail in the Estrella Mountains, the Maricopa Mountains, and the Gila Bend Mountains confirms what the songs

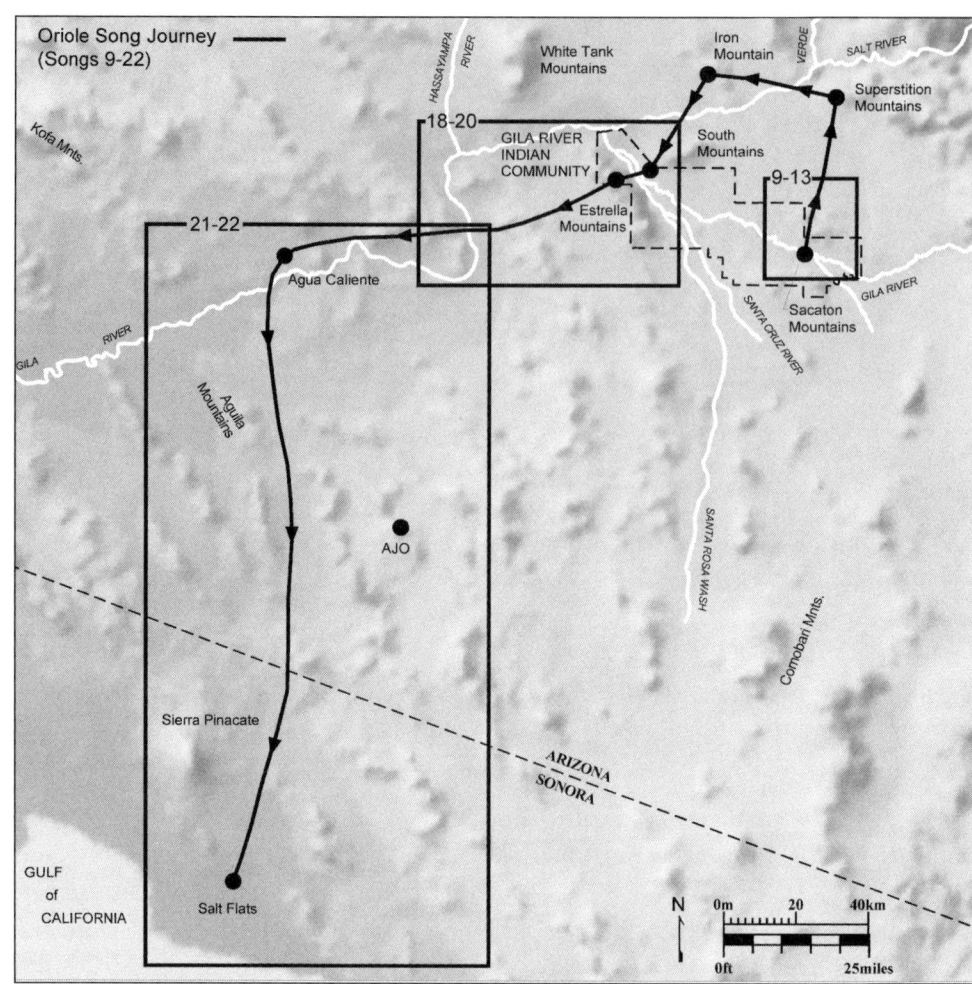

Figure 17.4. The Oriole song itinerary as shared by Vincent Joseph.

tell us—that this route stretched from the central Hohokam region in the Phoenix Basin to the western periphery of the Hohokam culture area. During the historic period, the trail linked O'odham territory with Yuman (Patayan) areas along the river, places where the Cocamaricopas, Opas (including O'odham and Opas living in mixed villages in the Gila Bend), and Maricopas lived. West of the Maricopa Mountains, the trails have survived better than elsewhere, so we can see where the main trail, known as the Komatke Trail, split into numerous smaller trails going to villages scattered along the Gila River valley.

Just as rock art and shrines appear along the trails through the Santan Mountains, we find them here, too, along with fragments of broken pots. Where the Komatke Trail enters the valley of the Gila Bend, archaeologists have found shell artifacts,

Figure 17.5. A portion of the Komatke Trail heading through the Maricopa Mountains.

including an unmodified *Glycymeris* shell. The presence of these shells, so prized among the Hohokam for making bracelets, indicates that people traveled and traded goods from the Gulf of California all the way to the middle of the Gila River drainage and beyond.

In the first part of the song journey, twenty songs detail the steps from Blackwater, on the Gila River reservation, to the Estrella Mountains—some ninety-five miles. Only two songs then carry us to the end of the journey, first to a cluster of hot springs, which some researchers identify as those at Aguacaliente, west of Gila Bend, and finally to the salt flats on the Sonoran coast.

The Tohono O'odham of southwestern Arizona traveled to this shore to gather salt. Along the northern beaches of the Gulf of California, high tides leave salt deposits in a largely uninhabited and waterless landscape. In summer, the Tohono O'odham made pilgrimages to the salt beds, following trails across the desert that connected the few hidden natural tanks holding rainwater.

Salt pilgrimages were significant religious events. Young men and experienced, older male leaders made the arduous journey, which in times past presented some of the same hardships and dangers as going to war. Salt pilgrimages also offered opportunities for dreaming and acquiring spiritual power. Once a man volunteered for his first salt journey, he was committed to three additional trips in successive years, and at the end of each trip, he was purified as if returning from battle.

The journey described by the Oriole song cycle describes a route unique to the Akimel O'odham (Gila River Pima). This route suggests that they may have practiced their own salt journey, which included travel to the hot springs in Aguacaliente before turning south to the salt flats in Mexico.

**Musical Rasps and Calendar Sticks**

For Vincent Joseph, the portion of the Oriole songs describing the westward journey ended at the salt flats. But the song series itself continues with many songs unrelated to the journey.

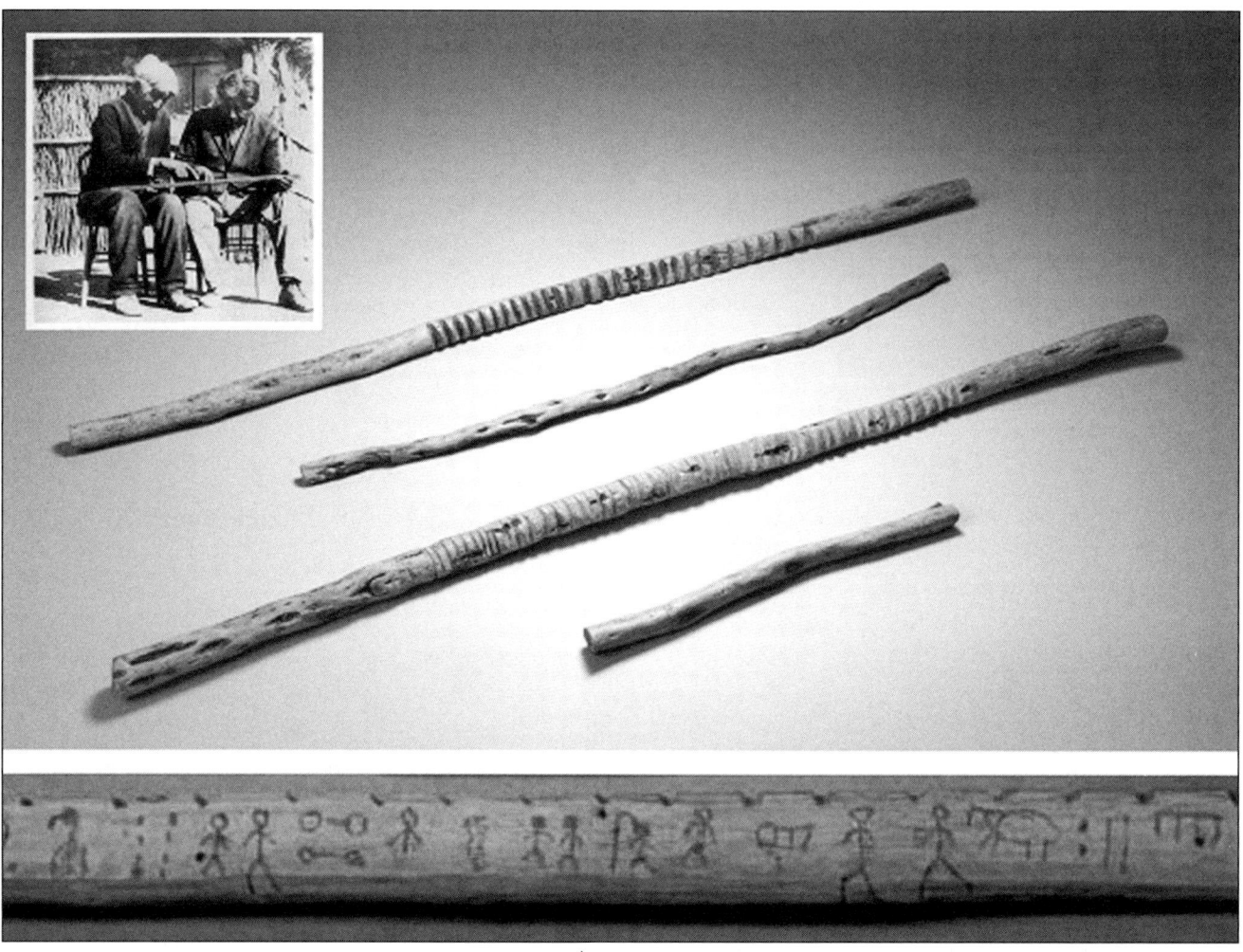

Figure 17.6. *Top:* two sets of rasping sticks used by the late George Kyyitan. *Top inset:* Joseph Head (Maricopa) reading a calendar stick to Henry Soalikee, 1921. *Bottom:* designs (o'ohadag) on a calendar stick.

In Joseph's rendition, the second-to-last song in the series concluded:

> And now we stop singing and scatter.
> Here on our seats our poor scraping sticks lie,
> With song-marks marked where they lie.

Scraping sticks (*hi:ʃkut*) are musical rasps used to accompany traditional singing. They consist of two lengths of greasewood or ironwood, one notched at regular intervals and scraped with the other stick in a rhythm appropriate to the music. People also play rasping sticks against a basket or gourd resonator. After repeated performances, the spiritual essence of the songs becomes part of the scraping sticks used to perform them. Even when the sticks are no longer used, they retain this spirituality and should be handled respectfully.

Some singers adorn their rasping sticks with carved or painted designs called *o'ohadag*, also known as "song flowers" (*ñe' ñei hiosig*). Bahr and his colleagues identify the word for song marks on the sticks as *o'ohon*, meaning "writing," but we feel that *o'ohadag* is more appropriate. O'ohadag are representations of the spiritual presence of the songs in these instruments, obtained through their use in performances. The designs are not strictly decorations but are emblematic of the singers' spiritual accomplishment, particularly the song journey. As the Oriole song suggests, song marks appear on the rasping sticks only when the performance is completed.

O'ohadag appear in many other places as well.

Figure 17.7. The Akimel O'odham musician Sides plays a rasping stick on a basket, 1919.

The images in rock art, commonly seen along trails and at the locations mentioned in song itineraries, are o'ohadag. O'ohadag are also found on O'odham calendar sticks (*hikanaba*), which in many ways resemble musical rasps. The designs on rasps document spiritual song journeys in geographic space, whereas calendar sticks provide a temporal or chronological itinerary—a time line—relating the present to the past. This is an important distinction between O'odham systems of geographical and historical reckoning.

O'odham calendar sticks typically are longer, straighter pieces of wood than rasps, but they, too, are notched at regular intervals, each notch representing a year. The calendar stick keeper places painted figures or incised and painted symbols alongside the notches as reminders of the events that distinguished that year from every other. Each calendar stick is unique to the calendar stick keeper, but the stick itself is like a page out of the shared history of all O'odham. The keeper might pass the stick and its historical narrative on to a successor, or it might be buried with him at his funeral. The historical narrative and the symbols for each event become part of the stick itself, just as songs become part of scraping sticks. Even after the keeper has died, the stick remains a repository of personal historical knowledge.

We know the histories told by only a few of the calendar sticks that have survived, when their accounts were written down in English. These histories generally span the mid-nineteenth to the early twentieth centuries and mainly record events of central interest to the O'odham, including battles with Apaches, deaths, famines, epidemics, and ceremonies. As objects, calendar sticks provide rare insights into O'odham historical reckoning. The excerpt presented in the sidebar, covering the years from 1888 to 1891, offers a glimpse into the events people sought to remember.

**Blackwater Calendar Stick Events**

1888–89. A Papago [Tohono O'odham] who knew the Bluebird series of songs sang for the Santan people during the festival held by them. The captain of the native police and the calendrist went to Fort McDowell with three other men to act as scouts for the soldiers stationed there. During the year, an epidemic carried away three prominent men at Blackwater.

1889–90. The wife of the head chief died.

1890–91. In the spring of 1891 occurred the last and most disastrous of the Gila floods. The Maricopa and Phoenix Railroad bridge was swept away, and the channels of both the Gila and Salt Rivers were changed in many places. The destruction of cultivated lands led to the change of the Salt River Pimas from the low bottoms to the mesas.

—Translated by Frank Russell, *The Pima Indians*

Descriptions of calendar stick narration also suggest that O'odham recited their histories in much the same way they sang song cycles. Just as songs had to be sung in the correct geographical sequence, historical narratives had to be told in the correct chronological order. O'odham songs and historical reckonings are performed in similar ways, but songs describe space and histories describe time. This distinction certainly did not escape

O'odham singers and calendar stick keepers.

Are the marks on calendar sticks similar to the song marks described for scraping sticks in the penultimate Oriole song? We think that this is likely. Historical recitations establish the present in reference to past events, in much the same way song journeys describe a spatial universe composed of interconnected spiritual places where things happen. Each provides the community with a path toward spatial and temporal awareness. Designs (o'ohadag) on rasps are abstract representations of songs and describe spiritual events associated with particular places. Their counterparts on calendar sticks represent historical events.

Figure 17.8. A song flower (ñe'ñei hiosig) on a rasping stick. Song flowers may appear as designs representing flowers, mountains, or other figures or as the notched cross shown here (center).

## Sacred Landscapes and History

What does all of this mean? We suggest that much of Native American infrastructure derives from traditional knowledge that relates ideas to facilities, such as trails, that allow societies to function. In the case of the O'odham, the retelling of song journeys generates a shared knowledge of O'odham geography, defined by significant places and the connections among them—by important sites and the spiritual events that happened there. Calendar stick histories explain the present in terms of past events. Known calendar stick histories begin in the nineteenth century, a time of substantial and sometimes catastrophic change in the O'odham world. Tribal histories describe in O'odham terms the cumulative effects of non-Indian incursions into their land and the disruption of their social relations and traditions.

How old are O'odham songscapes? Archaeological evidence reveals that the places the songs describe have been in use since the time of the Hohokam, centuries before Vincent Joseph's gift of the Oriole songs to anthropologists. Does the great age of these trails and sites mean that Hohokam people undertook spiritual, as well as practical, journeys similar to those of the O'odham? We do not know how the meanings of the paths and places may have changed for succeeding populations, but the fact that today's O'odham songscape trails connect these sacred sites of the past demonstrates their continued importance to the O'odham today.

Songscapes refer to the spatial and spiritual order of places and things, not to historical events. It is in this sense that they can be related to spiritual travels such as the salt pilgrimage. Through dreaming and song performance, O'odham singers raise their consciousness of the landscape.

What is the future of O'odham song? Modern development obscures traditional landscapes every day, yet songscapes continue to exist as new generations learn the songs from their elders. Without the songs, little would be left of the traditional O'odham worldview or of the sacred landscapes they portray. Working together, traditional community members and archaeologists may be able to recover ancient landscapes. O'odham trail archaeology, in particular, benefits enormously by taking into account the song journey described in the Oriole series. By learning about the traditional songscapes and the geographies they describe, archaeologists can hope to recover both the physical traces of past landscapes and their meanings for the people who created them. For the O'odham, their land persists in the songs and the songscapes they create. The final Oriole song, translated by Barnaby V. Lewis, says:

> The songs are ending as they go their separate ways.
> From the center of our songs, the wind comes,
> Flowing back and forth,
> Erasing the tracks of the people [preparing the ground for future sings],
> Ready to place them here again.

**J. Andrew Darling** earned his PhD in anthropology from the University of Michigan, Ann Arbor. He is a principal investigator and co-owner of Southwest Heritage Research, LLC, and the former director of the Gila River Indian Community Cultural Resources Management Program and Huhugam Heritage Center. He is the author of numerous publications on Southwest and Mesoamerican archaeology. He has worked in northern Mexico, Peru, Arizona, Colorado, and New Mexico and has expertise in tribal archaeology, traditional cultural properties, and international repatriation issues.

**Barnaby V. Lewis** is the Tribal Historic Preservation Officer for the Gila River Indian Community in charge of consultation with federal and state agencies and the archaeological community regarding repatriation and protection of affiliated human remains. He is an enrolled member of the Gila River Indian Community of southern Arizona and is well known for his cultural preservation efforts, particularly in keeping alive the songs of the Akimel O'odham.

*This chapter represents the views of its authors only. It does not in any way represent the views of the Gila River Indian Community, its members, or the O'odham of other communities in the United States or Mexico.*

Figure 18.1. Stairs on the mesa north of Garcia Canyon.

# Ancient Trails of the Pajarito Plateau

*James E. Snead*

In the late 1890s, the former governor of the Territory of New Mexico, LeBaron Bradford Prince, made a long-anticipated pack trip into the wilds of the Pajarito Plateau west of Santa Fe. In addition to being a political heavyweight, Prince was an avid student of the region's pre-Columbian inhabitants; a dusty collection of relics added flair to his fashionably Victorian home. On this occasion, Prince's excursion took him into what is now Bandelier National Monument to visit the Shrine of the Stone Lions, described nearly twenty years earlier by Adolph Bandelier. In the process, Prince found himself musing over another of the plateau's secrets, one about which Bandelier had said little. Ancient trails, worn deeply into the bedrock, led to and from villages abandoned before the Spaniards ever arrived in the Southwest (plate 17). "They are not 'footprints on the sands of time,'" Prince wrote in a small book published after his journey, "but in the rocks of eternity."

In the 1980s and 1990s, I was part of the archaeological teams from the National Park Service that comprehensively studied Bandelier National Monument, and we, too, used the old trails. Looking for a way to climb a difficult canyon slope, we sometimes hiked around an outcrop to spy a smooth set of steps leading upward, a path built in ancient times by Pueblos facing the same problem. We were following, quite literally, in their footsteps.

Ironically, until recently, we knew almost as little about these trails as Prince did. For generations, researchers working on the Pajarito Plateau concentrated on its residential pueblos as sources of information about the past. Digs at these sites yielded the traditional items of archaeological interest: pottery, stone tools, wooden beams for dating, and plant and animal remains pertaining to diet and farming. Trails, in contrast, seemed to offer little useful information.

In recent years, we have begun to pay attention to a picture bigger than that painted by our excavations. Because people of an earlier era spent most of their time outside their homes, the argument goes, shouldn't we be searching for evidence of what they did outdoors? And shouldn't we use this information to develop new interpretations of past lives? Embracing such an approach, we can see the archaeological record not as simply a matter of disconnected "sites" where things can be dug up, but as the combined traces of a whole cultural landscape—a place where topography, environment, and the countless, subtle features left behind by human action can be viewed together as a complex record of human history.

The isolated canyons of Bandelier offer an extraordinary proving ground for the cultural landscape approach. Visible but little noticed under the piñons and ponderosa pines are the fields the ancestral Pueblo people used, the small shelters they built to keep out of summer storms, the lookouts they maintained to watch for game and visitors from other villages, and the petroglyphs they placed to convey information to knowledgeable observers. Knitting this landscape together, giving it structure and meaning, is the faint but tangible network of ancient trails. In other parts of the Southwest, particularly at Chaco Canyon, archaeologists have

Figure 18.2. Ancestral Pueblo trail at Tsankawi, photographed in the early twentieth century.

Figure 18.3. Parallel paths worn into the tuff bedrock west of Tsankawi Pueblo.

identified a network of roads that was important to local society. Trails, if somewhat less dramatic, clearly played similar roles.

To explore this idea of cultural landscape, some colleagues and I launched the Pajarito Trails Project in the early 1990s. Building from a foundation laid in earlier research, I developed a standard recording procedure for trails and set out to look for them in Bandelier and other parts of the plateau. The vast scale of the region and the modest resources available meant that we set no deadline for completing the project; we do the work as time permits. Yet, with the support of the National Park Service, the Friends of Bandelier, and Los Alamos National Laboratories, we have already compiled a fascinating body of information about the Pajarito trails.

## Trail Structure

Interpreting the Pajarito trails involves, first, the deceptively simple process of documenting their characteristics. It seems simple because the volcanic geology of the Pajarito Plateau creates unique conditions for trail preservation. The same soft, tuff bedrock that allowed the ancient inhabitants to carve rooms into the sheer canyon walls can be eroded by the passage of human feet, meaning that the oldest trails are quite literally worn into the stone. Centuries of foot traffic have marked these routes deeply and unmistakably. In some places, two or even three separate trails run parallel to each other or are interwoven. Many of these trails show signs of having been constructed intentionally, with steps and other features carved into the rock by hand.

Figure 18.4. Petroglyph panel along trail and near stairs, Tsankawi Mesa.

Despite these advantages, determining the age of a trail can be a problem. In many places, trails made yesterday might look identical to those made centuries ago. I once spent an entire morning working with a field crew of the Bandelier Archeological Survey recording a trail marked by cairns of piled stones, only at last to meet the hiking stockbroker who was setting them up. True ancestral Pueblo trails, however, usually include features lacking in more recent examples. The Pueblos' concept of steps, for instance, was quite different from ours. Instead of wide, flat, American-style steps, theirs were the width of a single foot, and the climber negotiated them as if on a ladder. Most impressive are the elaborate sets of steps known as staircases, which often extend for dozens of feet up the steep canyon sides.

Particular trails are also marked by petroglyphs. We find these trail markers in various situations, but usually they appear on rock faces where people climbing up from below could easily have seen them. One trail near Tsankawi ascends the cliff face immediately next to a dramatic petroglyph of a male figure that clearly marks the route. Other trail markers also consist of humanlike figures and geometric designs. Although we do not understand the symbolic repertoire of the trail markers, it is tempting to see them as indicators of territory or perhaps local identity that people traveling through would have readily understood.

## The Trail System

My colleagues and I have recorded more than seven miles of trails on the Pajarito Plateau. This figure sums only the combined lengths of visible trail "segments" that we have observed and is thus a fraction even of those that are already known. Entire swaths of the Pajarito, particularly to the north, have never been surveyed for trails, and we think it likely that the original system stretched for hundreds

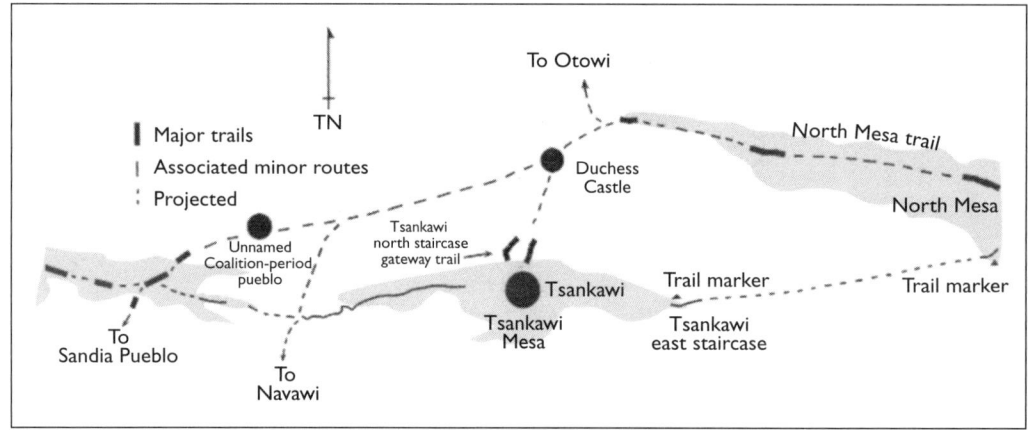

Figure 18.5. Route of the North Mesa Trail through the Tsankawi area.

of miles. Even with this small sample, we are increasingly able to understand the organization of the trails as they snake across the landscape.

With the available evidence, we divided the Pajarito trails into three categories: local trails, major trails, and trail networks. Local trails appear to have been used only by people who lived in their immediate vicinity. They were routes leading from mesa-top farms to sources of water in the canyons below, or paths to small hilltop shrines. Major trails linked people and places separated by greater distances, ultimately forming a system that spanned the entire Pajarito and connected it to surrounding regions. In some cases, major trails can still be followed for miles. In others, they are visible only when they cross a topographic obstacle such as a narrow, rocky ridge, where they are sometimes worn more than three feet deep. One trail in particular, the Old Pajarito Trail, may originally have run the entire length of the plateau, linking northern communities near Puye to those in the vicinity of Cochiti, dozens of miles to the south, and passing through many of the most thickly settled tracts en route.

Trail networks are more complex yet. They consist of trails or trail segments positioned so close together that they cannot really be considered separately. A trail network is most evident at Tsankawi, where we found seventy-four segments on a single mesa. The Tsankawi trails are bewilderingly complicated, linking the community house on the mesa top with cavate pueblos and other features on the flanks of the mesa and all of them to the valley below. This network was created by centuries of changing patterns of movement, a process ongoing even now. Some of the ancestral Pueblo trails at Tsankawi are part of the interpretive loop established by the National Park Service and show signs of recent wear, whereas adjacent segments that are not part of the system are disused and covered by pine duff.

Many of the major trails connect pueblo sites that date to the beginning of substantial ancestral Pueblo occupation of the plateau, probably before 1200 CE. These trails run east-west, from the Rio Grande to the Jemez Mountains, and north-south, traversing canyons to connect small settlements. Interestingly, people seem to have used these trails even after they abandoned their small settlements in favor of larger, later pueblos, which often lay some distance from the main routes. The North Mesa Trail, for instance, links several of the older sites, is deeply worn, and includes several constructed stairs. Although Tsankawi Pueblo lies directly south of this trail, most of the feeder routes that connect it to the major trail are relatively modest. The simplest explanation for this pattern is that the North Mesa Trail was established during the initial colonization of the plateau and continued in use even after many of the places it originally connected had been abandoned.

It is also apparent that the trail network saw use long after most Pajaritans had moved down into villages along the Rio Grande. During the historic era, the Pajarito served as a hunting ground, sacred retreat, and refuge for the Pueblos, who used the trails to travel across the mesas as their ancestors had before them. In several places, major trails are blocked by game traps—deep pits hacked into the bedrock to be used in hunting deer. Such traps would have been both inconvenient and dangerous at a time when the trails were used every day. The deer population must have been lower then, too, especially in residential areas. The game traps likely were made in more recent centuries, when deer were the only permanent residents of the plateau.

**Guard Pueblos**

As we gain a better understanding of the movement of people along the Pajarito trails, we begin to find answers to other archaeological questions. For instance, they may help us to better understand the way in which local communities were organized. In the historic period throughout the Southwest, it was important to pay close attention to people traveling along trails, to keep an eye on both friend and foe. In some cases, people built "guard pueblos" as watchposts from which to monitor traffic and protect the home village. The modern Hopi-Tewa community of Hano, in Arizona, began as just such a guard pueblo three hundred years ago.

The way several of the smaller Pajarito pueblos relate to the trail system suggests that they might have served as guard pueblos. A good example is a pueblo called Duchess Castle that sits in the valley north of Tsankawi Mesa. Duchess Castle is situated near the point where the North Mesa Trail would have crossed the valley from east to west. It also lies on the most logical route between Tsankawi and the Otowi community to the northwest. Because the flat valley bottom in the vicinity offers many good places for building a pueblo, the position of the site at the probable intersection of these trails is not just a coincidence. It is reasonable to imagine the residents of Duchess Castle guarding the junction on behalf of their kin on the mesa top to the south. Rainbow House in Frijoles Canyon is another good candidate for a guard pueblo, and there may well be others.

**Gateway Trails**

We often notice during our surveys that some of the Pajarito trails are far more complicated than was necessary to make travel easier. Sometimes the Pajaritans laboriously carved steps into relatively gentle slopes or established a parallel route even though an older route appears to have been quite adequate. I am increasingly convinced that the action of constructing the trail was as important as the trail itself and that the process was deeply symbolic.

The clearest example of this sort of labor investment comes from the half-dozen staircases recently recorded on the plateau. Each one has deeply incised steps, additional features such as handholds, and parallel and sometimes intertwined routes. In the Capulin Staircase in the southern part of Bandelier, several intertwining sets of stairs, each composed of dozens of steps, climb the face of the steep slope of Capulin Canyon. When taken in context with associated trail markers, this complex represents a remarkable effort when in fact just one of the parallel routes would have sufficed.

I suggest that the making of these elaborate staircases was an activity important in its own right. In Pueblo ritual, repetition is part of what makes an act sacred. Repeated cycles of song, dance, and prayer reaffirm relationships between people and the cosmos and serve as constant reminders of those ties. Some physical tasks, such as plastering the interior walls of ceremonial rooms and kivas, appear to be parts of repetitive rituals as well. Some ancestral Pueblo kivas excavated by archaeologists show evidence of continual replastering. One, at the village of Hawikuh, near Zuni, was replastered at least sixty-three times, and a kiva in Frijoles Canyon shows evidence of twenty replasterings. It appears that the plastering and repainting "renewed" the walls and the images on them, perhaps demonstrating the piety of those who took part. It seems possible that building other features had a similar purpose. Perhaps the periodic renewal of trails and stairs was a ritual act rather than merely construction to improve traffic flow.

It is also interesting that only certain sets of stairs were treated in this way, usually no more than one per community. I think that this reflects something akin to the building of entryways into medieval European towns: there were many ways in and out, but only one formal "gate" for important ceremonies and processions. Ancestral Pueblo people put their energy into stairs rather than gates, but they were responding to a similar impulse. Anyone coming up the trail from outside the community would have turned a corner and seen a magnificent stairway

Figure 18.6. Close-up view of a Tsankawi Mesa trail.

ahead, a sign of the spiritual propriety and strength of the people who lived there. I call these features gateway trails. Together with trail markers and nearby guard pueblos, they served notice to travelers from the outside that they were approaching a place with a strong history and identity and whose people were to be treated with caution and respect.

**Trails and "Human Space"**

The ancestral Pueblo trails of the Pajarito Plateau yield a surprising amount of information about the activities of the people who used them. They represent patterns of movement at different scales that literally tell us who went where and when. Putting that information together with other knowledge about the cultural landscape, we can reconstruct economic strategies and political organization. Above all, the trails indicate that networks of travel across the Pajarito were established early in the human history of the region and remained in use until very recent times. As we survey new areas, we will better understand the nature and extent of the system.

I also think it evident that trails were not simply functional. Ancestral Pueblo people went everywhere on foot, and the evidence suggests that their movement itself had symbolic connotations. Over time, as people walked the same paths to and from their villages, shrines, fields, neighbors, and hunting grounds, the routes took on a sacred character. Trails leading to Tsankawi, Yapashi, Puye, and Tyuonyi were invested with meaning that would have been of potent significance to those who walked along them. Periodically, the people built stairways to make that meaning clear.

All of this suggests that the archaeological study of trails is important for understanding how the Pueblos organized their world—how they created "human space" from nature. As the Pajarito Trails Project continues, we will get a firmer grasp on the

network of movement and interaction across the plateau. Perhaps the best summary of the potential that the trails offer to our understanding of the Pajarito Plateau comes from Bradford Prince, who wrote that they "tell vividly and more lastingly of the long occupation and vast numbers of people of those ancient ruins than could the most enduring monument."

**James E. Snead** was a member of the Bandelier Archaeological Survey and is an associate professor of anthropology at California State University, Northridge.

Figure 18.7. Tsankawi North Staircase, a gateway trail to Tsankawi Pueblo.

# Picture Credits

Abbreviations
- AMNH     American Museum of Natural History
- ASM     Arizona State Museum
- BNM     Bandelier National Monument
- CCAC     Crow Canyon Archaeological Center
- CRMP, GRIC     Cultural Resource Management Program, Gila River Indian Community
- MimPIDD     Mimbres Pottery Image Digital Database
- MIAC     Museum of Indian Arts and Culture
- MNA     Museum of Northern Arizona
- MNM     Museum of New Mexico
- MVNP     Mesa Verde National Park
- NAU     Northern Arizona University
- NMAI     National Museum of the American Indian
- NPS     National Park Service
- PMAE     Peabody Museum of Archaeology and Ethnology
- SAR     School for Advanced Research
- USU     Utah State University

Color section, after page <000>:
Plate 1, illustration by Victor O. Leshyk, based on 3-D lidar scan data provided by Western Mapping, Inc., and original photography by Boone, Belnap, and Downum; Plate 2, collections of NMAI, cat. no. 13/4512, courtesy Catherine S. Fowler, photographer; Plate 3, courtesy Linda Dufurrena, photographer; Plate 4, courtesy Polly Schaafsma, photographer; Plate 5, painting by Glenn Felch, courtesy CCAC; Plate 6, illustration by Victor O. Leshyk, based on original photography by Boone and Belnap and concept by Christian E. Downum; Plates 7, 16, 17, 21, 22, courtesy David Grant Noble, photographer; Plates 8 and 24, photos by George H. H. Huey, used with permission; Plate 9, *left*, cat. no. H/12758, photograph by Deborah Flynn Post; *center*, cat. no. H/5144; *right*, cat. no. H/3270, all courtesy AMNH Anthropology; Plate 10, photograph by Walter McQuarry, courtesy ASM; Plate 11, courtesy Karen Adams, photographer; Plate 12, MimPIDD 3641, Mattocks Ruin, collections of the Logan Museum of Anthropology, Beloit College, LMA no. 16123; Plate 13, courtesy USU Eastern Prehistoric Museum, cat. no. CEUM 8490; Plate 14, courtesy Bruce Hucko, photographer; Plate 15, photograph by Adriel Heisey, used with permission; Plates 18–20, photos by Dan Boone and Ryan Belnap, Bilby Research Center, NAU; Plate 23, photograph by E. J. Peiker, used with permission; Plate 25, photograph Peter J. Pilles Jr., used with permission.

Front matter: *Frontispiece*, Ancestral Pueblo human figures in red and white, Canyon de Chelly, courtesy Russ Bodner. Map 1, drawn by Molly O'Halloran, courtesy SAR.

Chapter One: Figs. 1.1, 1.3, 1.6, photos by Steven R. Simms, courtesy USU Department of Anthropology; Fig. 1.2, drawing by Steven R. Simms; Figs. 1.4, 1.5, 1.7, drawings by Eric Carlson, courtesy USU Department of Anthropology.

Chapter Two: Fig. 2.1, courtesy David Grant Noble, photographer; Fig. 2.2, diagram by Katrina Lasko, adapted from Fig. 4 in William C. Allen, "Present and Past Climate," in *Settlement and Subsistence along the Lower Chaco River*, edited by Charles A. Reher (University of New Mexico Press, 1977); Figs. 2.3, 2.6, courtesy R. Gwinn Vivian, photographer; Fig. 2.4, courtesy Museum of Indian Arts and Culture, Museum of New Mexico; Fig. 2.5, drawing by Katrina Lasko after Amy Elizabeth Gray in Stephen Plog, *Ancient Peoples of the American Southwest* (Thames and Hudson, 1997).

Chapter Three: Figs. 3.1–3.4, 3.6, 3.7, courtesy NPS, MVNP; Fig. 3.5, courtesy Bruce Hucko, photographer.

Chapter Four: Fig. 4.1, neg. no. S4-139872, courtesy University of Pennsylvania Museum; Fig. 4.2, plan by Neal Morris, courtesy CCAC; Figs. 4.3–4.5, courtesy David Grant Noble, photographer; Fig. 4.6, acc. nos. H-12981 (*left*) and H-13035 (*right*), photos by Bruce Hucko, courtesy AMNH; Fig. 4.7, drawings by Kristin A. Kuckelman, courtesy CCAC; Fig. 4.8, courtesy MVNP; Fig. 4.9, neg. no. 60720, courtesy Palace of the Governors (MNM/DCA).

Chapter Five: Fig. 5.1, neg. no. 28089, courtesy MNM; Figs. 5.2–5.4, courtesy NPS, BNM; Fig. 5.5, photograph by Anton Brkic, courtesy NPS, BNM; Fig. 5.6, adapted from original drawing by Kenneth Chapman, 1908, Laboratory of Anthropology Archives, MIAC; Fig. 5.7, illustration by Roy Ingraffia Jr., courtesy NPS, BNM; Fig. 5.8, courtesy NPS, BNM.

Chapter Six: Fig. 6.1, photograph by George A. Grant, courtesy NPS, Western Archaeological and Conservation Center, Museum Collection Repository Archives Program; Fig. 6.2, drawing by Patricia McCreery; Fig. 6.3, 6.5, courtesy David Grant Noble, photographer; Fig. 6.4, photograph by George A. Grant, 1929, courtesy NPS, Western Archaeological and Conservation Center, Museum Collection Repository Archives Program; Fig. 6.6, courtesy NPS, Chaco Culture National Historical Park, Museum Collection; Fig. 6.7, photograph by Paul Logsdon, courtesy Laurie Logsdon.

Chapter Seven: Fig. 7.1, photograph by Peter J. Pilles Jr., June 1982, courtesy Coconino National Forest; Fig. 7.2, illustration by Laurie Coveney-Thom, based on original 1934 drawing by J. C. Fisher Motz; Fig. 7.3, illustration by Victor O. Leshyk, based on 1999 map by James P. Holmlund, Western Mapping, Inc., Tucson, Arizona; Fig. 7.4, photograph by Cooper Aerial Photo, Inc., courtesy NPS; Fig. 7.5, photograph by Dan Boone and Ryan Belnap, Bilby Research Center, NAU; Fig. 7.6, courtesy MNA (NA-405NorthUnit.5).

Chapter Eight: Fig. 8.1, painting by Kia Gaspar, photograph by Addison Doty, cat. no. SAR.1979-1-4, courtesy SAR; Fig. 8.2, Bureau of American Ethnology, Annual Report XXIII; Fig. 8.3, photograph by Addison Doty, cat. no. IAF.P32, courtesy SAR; Fig. 8.4, photograph by Addison Doty, cat. no. IAF.P30, courtesy SAR. Fig. 8.5, photograph by Addison Doty, cat. no. IAF.C6, courtesy SAR.

Chapter Nine: Figs. 9.1, 9.4–9.6, courtesy David Grant Noble, photographer; Fig. 9.2, cat. nos. H/4375, H/4371, H/4379, H/4378, H/4185, H/4433, collections of AMNH Anthropology; Fig. 9.3, courtesy R. Gwinn Vivian, photographer; Fig. 9.7, photograph by Paul Logsdon, courtesy Laurie Logsdon.

Chapter Ten: Figs. 10.1, 10.4, 10.5, courtesy David Grant Noble, photographer; Fig. 10.2, courtesy Irene Silentman, photographer; Figs. 10.3, 10.6, photograph by Paul Logsdon, courtesy Laurie Logsdon.

Chapter Eleven: Fig. 11.1, photograph by Philip Johnston, courtesy MNA; Fig. 11.2, map by Alexa Roberts; Figs. 11.4, 11.5, courtesy MNA; Fig. 11.6, photograph by Jim Brewer, courtesy MNA; Fig. 11.7, courtesy Alexa Roberts, photographer.

Chapter Twelve: Fig. 12.1, map by Molly O'Halloran; Fig. 12.2, courtesy University of Utah Marriott Library Special Collections; Figs. 12.3, 12.5, courtesy Museum of Peoples and Cultures, BYU; Fig. 12.4, 29-5-10/A6766.1, © 2007 Peabody Museum, Harvard University; Fig. 12.6, courtesy Polly Schaafsma, photographer.

Chapter Thirteen: Figs. 13.1, 13.5, 13.6, 13.9, photos by Helga Teiwes, courtesy ASM; Fig. 13.2, Plate 43 in J. W. Fewkes, *Casa Grande, Arizona*, 28th Annual Report of the BAE, 1906–1907 (GPO, 1912); Fig. 13.3, map facing p. 12 in Omar Turney, *Prehistoric Irrigation in Arizona* (Arizona State Historian, 1929); Fig. 13.4, courtesy ASM; Fig. 13.7, courtesy Matts Myhrman, photographer; Fig. 13.8, photograph by Charles Miksicek, courtesy Suzanne K. Fish.

Chapter Fourteen: Fig. 14.1, photograph by Edward S. Curtis, about 1906, courtesy NAU Cline Library, NAU.PH.93.38.24; Figs. 14.2–14.4, 14.5 *top right*, 14.6, and 14.7, photos by Dan Boone and Ryan Belnap, Bilby Research Center, NAU; Fig. 14.5 *left*, courtesy Christian E. Downum; *center right*, courtesy Arizona State Museum (ASM 1415); *bottom*, courtesy Christian E. Downum, photographer, June 1994.

Chapter Fifteen: Figs. 15.1 MimPIDD 2091, PMAE 95787, 15.3 *left*, MimPIDD 2475, PMAE 95960, 15.5 *top left*, MimPIDD 2531, PMAE 96194, 5.6 *bottom*, MimPIDD 2695, PMAE 94755, and 15.8, MimPIDD 9552, PMAE 94632, all from Swarts Ruin, collections of the PMAE, used with permission of the President and Fellows of Harvard College; Figs. 15.2 *left*, MimPIDD 8118, Mimbres Valley, cat. no. 53, 15.4 *left*, MimPIDD 8027, Wind Mountain, cat. no. WM2444, and 15.5, *bottom left*, MimPIDD 9654, provenience unknown, cat. no. 7009, all photos by Steven LeBlanc, courtesy Amerind Foundation, Inc., Dragoon, Arizona; Fig. 15.2 *right*, ASM C-64854, photograph by Jannelle Weakly, courtesy ASM; Figs. 15.3 *right*, MimPIDD 1595, provenience unknown, UC 3268, 15.5 *top right*, MimPIDD 1607, Eby site, UC 3093, 15.6 top, MimPIDD 1601, Eby site, UC 3127, and 15.6 *middle*, MimPIDD 1602, provenience unknown, UC 3264, all collections of University of Colorado Museum, used with permission; Fig. 15.4 *right*, MimPIDD 33, Goforth site, collections of ASM, no. GP6550; Fig. 15.5 *bottom right*, MimPIDD 7599, NAN Ranch Ruin, used with permission of Western New Mexico University Museum; Fig. 15.7, illustrations by Will Russell.

Chapter Sixteen: Figs. 16.1, 16.9, courtesy Russ Bodner, photographer; Figs. 16.2–16.8, 16.10, 16.11, 16.14, 16.15, courtesy David Grant Noble, photographer; Figs. 16.12, 16.13, photos by Karl Kernberger, courtesy Carolyn Kernberger.

Chapter Seventeen: Figs. 17.1, 17.5, courtesy J. Andrew Darling, photographer; Fig.17.2, map by Lynn Simon, courtesy of CRMP, GRIC; Fig. 17.3, courtesy Barnaby V. Lewis, photographer; Fig. 17.4, map by Lynn Simon, courtesy CRM, GRIC, drawn from Donald Bahr, Lloyd Paul, and Vincent Joseph, *Arts and Orioles: Showing the Art of Pima Poetry* (University of Utah Press, 1997); Fig.17.6, upper inset photograph by Edward H. Davis, no. N24596, courtesy NMAI; photograph of rasping sticks by Melissa Altamirano, courtesy CRMP, GRIC; photograph of calendar stick by Josh Roffler, courtesy CRMP, GRIC; Fig.17.7, photograph by Edward H. Davis, courtesy NMAI, no. N34324; Fig. 17.8, photograph by Melissa Altamirano, courtesy CRMP, GRIC.

Chapter Eighteen: Fig. 18.1, 18.7, courtesy James Snead, photographer; Fig. 18.2, photograph by George Bean, courtesy Denver Public Library, Western History Collection, no. P1596; Fig. 18.3, courtesy Robert P. Powers, photographer; Figs. 18.4, 18.6, courtesy David Grant Noble, photographer; Fig. 18.5, map by Molly O'Halloran; *page 165*, glyph taken from figure 16.9.

Other illustrations: *Page 62*, pot pattern drawing by Victor O. Leshyk; *page 63*, complex abstract life form petroglyph from Chavez Pass. From illustration by Jane Kolber. Courtesy Kolber. Re-drawn by Victor O. Leshyk.

# Chapter Credits

Chapter One "Making a Living in the Desert West" by Steven R. Simms originally appeared as ch. 13 in *The Great Basin* (pp. 94–103), 2008.

Chapter Two "Puebloan Farmers of the Chacoan World" by R. Gwinn Vivian originally appeared as ch. 2 in *In Search of Chaco* (pp. 7–13), 2004.

Chapter Three "Through the Looking Glass: The Environment of the Ancient Mesa Verdeans" by Karen R. Adams originally appeared as ch. 1 in *The Mesa Verde World* (pp. xviii–7), 2006.

Chapter Four "Ancient Violence in the Mesa Verde Region" by Kristin A. Kuckelman originally appeared as ch. 16 in *The Mesa Verde World* (pp. 127–135), 2006.

Chapter Five "Carved in the Cliffs: The Cavate Pueblos of Frijoles Canyon" by Angelyn Bass originally appeared as ch. 11 in *The Peopling of Bandelier* (pp. 86–93), 2005.

Chapter Six "Architecture: The Central Matter of Chaco Canyon" by Stephen H. Lekson originally appeared as ch. 4 in *In Search of Chaco* (pp. 22–31), 2004.

Chapter Seven "Wupatki Pueblo: Red House in Black Sand" by Christian E. Downum, Ellen Brennan, and James P. Holmlund originally appeared as ch. 11 in *Hisat'sinom* (pp. 78–86), 2012.

Chapter Eight "Zuni Religion and Philosophy" by Edmund J. Ladd originally appeared in *Zuni El Morro: Past & Present* (pp. 26–31), *Exploration*, 1983.

Chapter Nine "Yupköyvi: The Hopi Story of Chaco Canyon" by Leigh Kuwanwisiwma originally appeared as ch. 6 in *In Search of Chaco* (pp. 41–47), 2004.

Chapter Ten "Canyon de Chelly: A Navajo View," An Interview of Mrs. Mae Thompson by Irene Silentman, originally appeared in *Tsé Yaa Kin/Houses Beneath the Rock: Canyon de Chelly and Navajo National Monument* (pp. 50–56), *Exploration*, 1986.

Chapter Eleven "The Wupatki Navajos: An Historical Sketch" by Alexa Roberts originally appeared in *Wupatki and Walnut Canyon: New Perspectives on History, Prehistory, Rock Art* (pp. 28–33), *Exploration*, 1987.

Chapter Twelve "The Enigmatic Fremont" by Joel C. Janetski originally appeared as ch. 14 in *The Great Basin* (pp. 104–115), 2008.

Chapter Thirteen "The Hohokam Millennium" by Suzanne K. Fish and Paul R. Fish originally appeared as ch. 1 in *The Hohokam Millennium* (pp. 1–11), 2008.

Chapter Fourteen "Pottery of the Sierra Sin Agua" by Kelly Hays Gilpin and Christian E. Downum originally appeared as ch. 17 in *Hisat'sinom* (pp. 124–131), 2012.

Chapter Fifteen "Expressions in Black and White" by Michelle Hegmon originally appeared as ch. 8 in *Mimbres Lives and Landscapes* (pp. 64–73), 2010.

Chapter Sixteen "Anasazi Rock Art in Tsegi Canyon and Canyon de Chelly: A View Behind the Image" by Polly Schaafsma originally appeared in *Tsé Yaa Kin/Houses Beneath the Rock: Canyon de Chelly and Navajo National Monument* (pp. 24–33), *Exploration*, 1986.

Chapter Seventeen "Songscapes and Calendar Sticks" by J. Andrew Darling and Barnaby V. Lewis originally appeared as ch. 16 in *The Hohokam Millennium* (pp. 130–139), 2008.

Chapter Eighteen "Ancient Trails of the Pajarito Plateau" by James E. Snead originally appeared as ch. 10 in *The Peopling of Bandelier* (pp. 78–85), 2005.

# Index

Numbers printed in *italics* refer to illustrations; numbers beginning with uppercase P refer to plates; numbers printed in **bold** refer to maps.

Adams, Karen R., 19–25
Agave, *114*
Ahil, Juanita, P10
Akimel O'odham, 153
Alameda Brown Ware, *119*, 119–120, 121, 125, P6
American Museum of Natural History, 45
Amphitheater, Wupatki Pueblo, *54*, 59
Amsden, Charles, 137
Anasazi. *See* Ancestral Pueblo
Ancestral Pueblo: burials, 21, 22; culture area, **ix**; houses, 46, 48; material goods, 83; Navajo oral tradition about, 82–84; rock art, 72, 76, 77, *139*, *140*, *142*, *143*, *144*, 145–146; trails, *160*; violence, 27
Animals: classification by habitat, 132; domesticated, **33**, 34; on Hohokam pottery, *128*; hunted, *xii*, 1, 1–3, *6*, 6–7, 24, *24*, 100–101, 163, P2; on Mimbres pottery, *126*, 128, *129*, *130*, 130–133, *131*, *132*, *134*, *135*, P12; nondietary uses, 2, 24; on rock art, 139, *139*, 140, 146, 147, P4; songs composed by spirits of, 150; Sonoran Desert, 114; as totems, P20
Anthropogenic ecology, 24–25
Anthropomorphs in rock art: ancestral Pueblo, *140*; Basketmaker II period, *136*, 137–138, *138*; Canyon del Muerto, P22; negative images, 146
Anthropophagy, 33
*Ants and Orioles: Showing the Art of Pima Poetry* (Bahr), 149, 151
Aphid honey, 7–8
Archaeology: Chaco Project, 46–47, 48; Crow Canyon Archaeological Center, 27; cultural landscape approach, 159–160, 163–165; early ideas about Chaco Canyon, 45; evidence of Hopi in Chaco Canyon, 73, 74; Fremont, 97, 98; Hohokam, *116*, 116–117; Pajarito Trails Project, 160; School of American Archaeology, 43; Snake Kiva, 39; Solstice Project, 51; study of pottery, 129; Wupatki Pueblo, *61*
Archaic period: Chaco Canyon, 12; Great Basin basketry, 4–5, *5*; nomadic routes established, 4
Architecture: Chaco, 46, 47, 47–48; diverse styles at Wupatki, 60; Fremont, 98–100, *99*, *100*, 104; Pueblo style (historic), 52. *See* specific types of structures and specific structures
Atlatl Cave pictographs, Chaco Canyon, *72*
Aztec Ruins, 51–52, *78*
Aztec West, *53*

Bahr, Donald, 149, 151, 154
Ballcourts: Hohokam, 110, *111*, 112, 115; Wupatki Pueblo, *54*, 59
Bandelier, Adolph, 41, 42, 159
Bandelier National Monument, 159
Bandelier Tuff, 38
Bannister, Bryant, 46–47
Barrier Canyon rock art, 138–139, P4
Basketmaker II period: anthropomorph rock art, *136*, 137–138, *138*; bird artifacts, 142; birds in rock art, *140*, 140–143, *143*; influences on rock art, 138–139; violence, 27
Basketmakers: in Chaco region, 72, 73, 74; migrations, 75–78; pictographs in Chaco region, *72*

Basketry, 4–5, *5*, 101
Bass, Angelyn, 37–43
Beans, 21
Betatakin rock art, 146
Big Fire Curing Society, 69
Bighorn sheep, *6*, 6–7
Bilk Creek Mountains, P3
Birds: in Basketmaker II rock art, *140*, 140–143, *143*; in burials, 142; spiritual aspects, 141–142
Blackhorse, Taft, 51
Blow holes, 60
Blue Star Katsina, 76
Bonito phase (Mesa Verde), 46
Bow Priesthood, 66
Bowyer, Vandy, 24–25
Breathway, Zuni, 65
Brennan, Ellen, 55–62
Brewer, Jimmy, 92
Brewer, Sally, 92
Brigham Young University, 98
Brody, J.J., 42, 127, 130, 133
Built environment: Chaco and Mesoamerica, 51; changes during historic period, 19; Great Basin, 2–4; Mesa Verde, 24–25; terracing for farming, 20; water management, 14, *15*, 21, 98, **107–108**, 110, 112, 115, *116*. *See also* Trails
Burials: ancestral Pueblo, 21, 22; birds in, 142; Fremont, 98, 104; Mimbres, 128–129; at Wupatki Pueblo, 60; Zuni, 66
Bustos Wickiup Village, 3–4

Caching practices, 3–4
Cahokia, 52, *53*
Calendar sticks, *154*, 155–156
Canals: Chaco Canyon, 14, *15*; Hohokam, **107–108**, 110, 112, 115, *116*
Cannibalism, 33
Canyon de Chelley, *83*, *85*, 137; Coconino Navajo, 90; Navajo oral tradition, 81–87; trade with ancestral Pueblo, 73–74; White House Ruins, *80*
*Canyon de Chelley: Its People and Rock Art* (Grant), 143
Canyon de Chelley rock art, *84*, *138*, *142*; ancestral Pueblo, *139*, *143*, *145*; Basketmaker anthropomorphs, *136*; Basketmaker II, 138–139, *140*, 140–143, *143*; post–Basketmaker II efflorescence, 142–143
Canyon del Muerto, *85*; Ceremonial Cave, P22; Mummy Cave, P21; rock art, *84*, *138*, *139*
Capulin Staircase, 163
Carleton, James, 89
Carson, Christopher "Kit," 89
Casa Grande National Monument, *107*
Casa Rinconada, *79*, 79
Castle Rock Pueblo, 27, 32, *32*, 35
Cavate pueblos of Frijoles Canyon, *38*; architectural finishes and embellishments, 37, 39–40, *40*; construction, 37–39, *39*, 42–43; kivas, 30, 40, *41*, 41–42; modern history, 43; number, 37, 40; preservation, 43; relationship with freestanding masonry pueblos, 37, 38; uses, 40–43, *41*, *42*
Cavate pueblos of Pajarito Plateau, 37, 162
Cave 7, Whisker's Draw, *26*, 27, 32
Ceramics. *See* Pottery

Ceremonial Cave, P22
Ceremonial structures: amphitheater, 54, 59; ballcourts, 54, 59, 110, *111*, 112, 115; kivas, 39, 40, *41*, 41–42, 48, 49, *78*, 78–79, *79*, 163, P8
Ceremonial wands (Pueblo Bonito), 74
Chaco Canyon region, P15; architecture, 46, *47*, 47–48; Basketmakers, 72, 73; Chaco Wash, *10*, 13, *16*, P15, P24; downtown Chaco, 14, 50–51, P15; farming, 11–16, *14*; great houses outside Chaco Canyon, 16, 51, 53, 60; Hopi oral tradition, 73–74, *74*; line-of-sight communication, 51; Native Americans cultural affiliation, 75; precipitation and temperature, 11–12, *12*, 13–14, 15–16; Pueblo del Arroyo, 48; roads, 51, *52*; rock art, 72, *76*, *77*; social strategies, 14–16; and Wupatki Pueblo, 60. *See also* Great houses in Chaco Canyon; Pueblo Bonito
*Chaco Meridian, The* (Lekson), 52
Chaco Project, 46–47, 48
Chaco Wash, *10*, 13, *16*, P15, P24
Chacra Mesa, 11, *12*
Chapman, Kenneth, *36*, 37, 41–42
Chetro Ketl, 48
Chimney Rock Pueblo, 51
Clans: Hopi, 73, 74, 75, 77; Zuni, 65, 66, 146
Clay miniature objects, 69, 101–102
Cliff dwellings, 29, *30*, P14. *See also* Mesa Verde region
Cliff Palace, construction, 45
Climate: corn-growing degree day units, 21; Hohokam culture area, 114; Mesa Verde region, 21–22, 34; modern Mesa Verde, 20; shift during late 1200s to early 1300s, 34; Sonoran Desert, 115; Wupatki, 61; Wupatki Pueblo, 55; Zuni rituals for, 67
Clothing, Fremont, 102, *102*, 105, *105*
Coconino Navajo, 89–91
Coffee-bean appliqué technique, 101
Cohonina, 58, *120*, 120–121, 124
Cole, Sally, 104
Colorado Plateau rock art, 138
Colton, Harold S., 58, 92, 93
Communication: line-of-sight, in Chaco Canyon region, 51. *See also* Trade; Trails
Confluence site, 98
Coolidge, Calvin, 58
Cordell, Linda, 14–15
Corn, 21, *22*; few depictions on pottery, 130; and Fremont, 100, 105; growing requirements, 12, 20, 21; indigenous varieties, P11; overdependence on, 35; as religious offering, 68, 69; storage, 22
Corn-growing degree day (CGDD) units, 21
Corrugated pottery, 123, *123*
Cosmology. *See* Religion
Cremation burials, 60
Crops: amount needed to sustain family, 21; beans, *75*; complete protein, 21; current Navajo, 82; Fremont, 97, 100; Hohokam, 115; squash, 21. *See also* Corn
Crow Canyon Archaeological Center, 27
Culture: areas, ix; classification systems, 132; and environment, 97, 159–160, 163–165
Curing societies, 66, 69
Cushing, Frank Hamilton, 107

Dams, 12, *13*, 14, *15*, 98
Dance courts, 59
Darling, J. Andrew, 149–157
Deadman's Black-on-red Ware, 123
Deadman's Fugitive Red Ware, 121
Deaths: and abandonment of homes, 92–93; caused by violence, 21, 22; feeding of deceased, 66, 67; scenes on pottery, 133; and transformation, 133. *See also* Burials
Deforestation, 61
Diet. *See* Food
Diné. *See* Navajo
Disease and population clustering, 34
Dismemberment, 33
Diversion dams, 14, *15*
Dogs, 24
Douglass, Amy, 124
Downtown Chaco, 14, 50–51, P15
Downum, Christian E., 55–62, 119–125
Driver, Jonathan, 24
Droughts: Chaco Canyon region, 16; Mesa Verde region, 21–22, 34; Wupatki Pueblo, 61
Duncan, Isadora, 134
Dutchess Castle, 163

Eagle Nest, Ute Mountain Tribal Park, *30*
Effigies, 147
Elsinore site, 98
Environment: and culture, 97, 159–160, 163–165; humans as part of nature, 8–9; Mimbres culture area, 133; modern Mesa Verdean, 20; sacred sites/landscapes, 156; Sonoran Desert, 114–115; use of micro-, 14
Extended burials, 60

Fajada Butte, *10*
Farming: Chaco Canyon, 11–16, *14*; competition for land, 61; Fremont, 98, 105; harvest festival, *64*; Hohokam, 113–114, 115; Mesa Verde conditions, 19, 21–22, 25, 34; modern, in Mesa Verde region, 25; overuse of fields, 61–62; sand dune method, 75; and Sunset Crater eruption, 58; terracing, *20*. *See also* Corn; Crops; Irrigation
Fenn cache, 4
Festivals, 103. *See also* Religion
Fewkes, Jesse Walter, 60
Figurines, 101–102, P20
Fire clouds, 120, 125
Fish, Paul R., 107–117
Fish, Suzanne K., 107–117
Fishing, 7, *8*, 24, 101
Five Finger Ridge site, 104, *105*
Flagstaff Black-on-white Ware, 122
Flexed burials, 60
Flute Player Cave, 144, *144*
Flute player motif, *143*, 143–145, *145*, 146
Food: animals eaten, 6–7, 114; for deceased, 66, 67; fat content, 7; Fremont, 100–101; historic Navajo, 85; Mesa Verde region native plants, 22–23, *23*, 24; on Mimbres pottery, 130, 131; overdependence on one resource, 35; prayer meals, 68, 69; preparation methods, 4–6, *5*, 7; religious fasting periods, 68, 69; as religious offerings, 68; social adaptations to shortages, 14–16; Sonoran Desert native plants, *113*, 114, *114*; sweets, 7–8; types gathered, 4–6, 8, 22–23, *23*, 24, 101, P10; violence caused by scarcity, 33–34. *See also* Corn
Food storage: caches, 3–4; cavates, 40, *41*, *42*; jars, 22; by modern Hopi, 74; pits, 100, 115; rooms, *31*, 50, *50*, 99, *99*, 100, 104
Fort Sumner, 86–87, 89–90
Fowler, Andrew, 52
Freeman, Katy, 51
Fremont: burials, 98, 104; clothing and adornments, 102, *102*, *103*, 105, *105*; culture area, **ix**, **96**; environment, 97; excavations, 97; food, 100–101; overview of culture, 98; pottery and basketry, 101–102, P13; rock art, 104–105, *105*; social life, 102–103; warfare, 104

Friedman, Richard, 51
Frijoles Canyon, 37–43, P16
Fritz, John, 50–51

Galaz Ruin, 132
Gambling, 102–103
Gaspar, Kia, 64
Gateway trails, 163–164
Geographical space, interpreting, 149
Gila River Indian Community, *148*, 149, *152*
Gillespie, William, 12
Gladwin, Harold, 116–117
Goodman Point Pueblo, 27, 28
Granaries, *31*, 99, *99*, 100, 104
Grand Gulch area cliff house, P14
Grant, Campbell, 143
Grasshoppers, 7
Gray-ware ceramics, 101
Great Basin, built environment, 2–4
Great Drought, 34
Great houses, Chaco Canyon outliers, 16, 51, 53, 60
Great houses in Chaco Canyon: architectural heirs, 52; connecting roads, 51; construction, 45, 49–50, 53; early, 15; as farming adaptation, 14, 15; outliers, 16, 51, 53, 60; peak, 47; precipitation effects, 16; room-to-kiva ratio, 49; size, 48; usage, 48, 49–50, 51; viewed as independent settlements, 45
*Great Pueblo Architecture of Chaco Canyon* (Lekson), 46–47
Great Pueblo period, 45
Guard pueblos, 163
Guernsey, Samuel, 138

Handprints in rock art, 138, *139*, *145*, 146, *146*, 147
Harvest festival, 64
Haury, Emil, *116*, 117
Havasupais, 60
Hayes, Alden, 46, 47, 51
Hays-Gilpin, Kelley, 119–125
Head, Joseph, *154*
Headdresses, Fremont, 102, 105
Hegmon, Michelle, 127–135
Hewett, Edgar Lee: cavate kivas, 41; on Chaco Canyon, 11; experiential learning, 43; Snake Kiva excavation, 39; Talus House foundations, 37
Hidden House Ruin, 122
Hisat'sinom: Cohonina, 58, *120*, 120–121, 124; pottery, *119*, 119–120, 121–123, *123*, P6; projectile points, P18
Historic period: changes in built environment, 19; languages in Hohokam culture area, 111; Pueblo architecture, 52; Pueblo religion, 141–142; rock art about Spanish, *84*; Shoshone hunting, 1; tool making, 4; trails, 162–163; trapping during, 162–163; US government campaigns against Navajo, 84–87
Hohokam: archaeology, *116*, 116–117; ballcourts, 110, *111*, 112, 115; burials, 60; canal systems, **107–108**, 110, 112, 115, *116*; Casa Grande National Monument, *107*; chronology, 115–116; Classic period, 58; community organization, 112–113, 115, 116; cultural hallmarks, 115; culture area, ix; designed shells, *110*; environment, 114–115; identity and influence, 110–111; Mesoamerican influence, 113; molded faces, *106*; mounds, 110, 112, 115; and O'odham songscapes, 151–152; pottery, 111, 115, 124, 128, *128*; settlement stability, 113–114, 115; shell jewelry, 60; stone palettes, *112*; trails, 153–155, 156
Holmlund, James P., 55–62
Hopi: clans, 73, 74, 75; food storage, 74; Kayenta ancestors, 58, 121, 122–123, *123*, P6; migrations, 60, 75–78; names for Wupatki Pueblo, 58; oral tradition about Chaco Canyon, 73–74, *74*; oral tradition of attack on Castle Rock Pueblo, 32; oral traditions about Wupatki Pueblo, 59, 60, 61; place of emergence, 60; pottery, *118*, P6; relationship with researchers, 74–75. *See also* Hisat'sinom
Hopi-Tewa, 118
Hosta Butte phase (Mesa Verde), 46
House blessing ceremony, 67–68
Housing: ancestral Pueblo pit, 48; Fremont pit, 98–100, *100*, 105; Hohokam pit, 115; and Navajo death customs, 92–93; pit houses and population estimates, 48–49; seasonal, 4, 82, 99–100; wickiups as temporary, *2*, 2, 3–4, 99–100. *See also* Cavate pueblos of Frijoles Canyon
Hovenweep Castle, *29*
Howard, Jerry, 112
Huérfano Mountain repeater station, 51
Huhugam. *See* Hohokam
Humboldt Cave, 4
Hunting: animals eaten, 6–7, 24, *24*, 100–101; methods, xii, 1, 1–3, 6, 6–7, 101, P2; patterns, 24; tools, 31, *32*, P19; trapping during historic period, 162–163
Huntington Canyon site, 104
Hwééldi, 86–87
Hyde Exploring Expedition, *26*, 27

Insects as food, 7
Irrigation: diversion dams, 14, *15*; Fremont dams, 98; Hohokam canal systems, **107–108**, 110, 112, 115, *116*; pot, 21; sand-dune dams, 12, *13*

Jackson, William Henry, 11
Janetski, Joel C., 97–105
Jennings, Jesse, 97
Jewelry, 60, 102, *103*
Johnson, Eleanor, *36*
Johnston, Philip, *91*, 92, 93, 94
Johnston, William R., 89, *91*, 91–92
Joseph, Vincent, 149, 151, 153–154
Judd, Neil, 97

Katsinas, *70*, 76
Kayenta ancestral Pueblo: population clustering, 58; pottery, 121, 122–123, *123*, P6
Kelly, Isabel, 4
Kidder, Alfred Vincent, 138
Kiet Siel, *140*
Kill holes, 128–129
Kin Kletso, 48
*Kin Kletso: A Pueblo III Community in Chaco Canyon* (Gordon Vivan and Mathews), 46
Kinship: Hopi clans, 73, 74, 75, 77; and living adaptations in Chaco Canyon, 14; and nomadic structure, 3; Zuni clans, 65, 66, 146
Kintigh, Keith, 52
Kivas (architectural structures): Aztec Ruins, *78*; Casa Rinconada, *79*, 79; cavate, 30, 40, *41*, 41–42; Chaco great house ratio of rooms-to-, 49; modern Rio Grande, 48; plastering as ritual, 163; Pueblo Bonito, 78–79, P8; purpose of ancient, 48; Yellow Jacket great house ratio of rooms-to-, 48
Kivas (Zuni men's societies), 65, 66
Kohler, Timothy A., 21, 24
Komatke Trail, 152–153, *153*
*Koyemshi*, 67, 67–68
Koyongtupqa, 73–74. *See also* Canyon de Chelley
Kuckelman, Kristin A., 27–35
Kulow, Stephanie, 129, 132
Kuwanwisiwma, Leigh J., 73–79

Ladd, Edmund J., 65–71
Lakeside Cave, 7
Languages, 8–9, 110, 111
Leadership, Zuni, 66
LeBlanc, Steven, 129–130
Lee, Charles W., 101
Lekson, Stephen H., 14, 15, 45–53
Lewis, Barnaby V., 149–157, 156
Life road/lifeway, Zuni, 65
Little Box Canyon Pueblo, P1
Little Colorado White Ware, 124, *124*
Local trails, 162
Long House (Frijoles Canyon), 37
Lovelock Cave, Nevada, 3

Macaws, 133
Maize. *See* Corn
Major trails, 162
Mantles Cave site, 102, 105
Masaw (Hopi spiritual guardian), 73, 75–78
Masonry: Pueblo Bonito, 47, *47*; Wupatki Pueblo, 58
Massing, 49–50
Mathews, Meredith, 24
Mathews, Tom, 46, 47
McElmo phase (Chaco), 46, 50
Mealing bins, 40, *41*
Men's societies, Zuni: initiation, 66, 67; membership, 65; participation in rituals, 66, 68
Mesa Verde National Park, 29
Mesa Verde region, *18*; Cave 7, *26*, 27; Cliff Palace, 45; construction time period, 45; defensive measures, 27–29, *28*, *29*, 30, 31, P14, P21; environments and climate, 20–22, 34; farming, 19, 21–22, 25, 34; migration to, 76; modern farming in, 25; native plant foods, 22–23, *23*, 24; non-Pueblo peoples, 34; population, 21–22, 28–29; Yellow Jacket kivas, 48
Mesoamerica: built environment, 51; and Chaco architecture, 46; influence on Hohokam, 113; influence on people of Sierra Sin Agua, *122*; trade with Mimbres, 133
Metates, 40, *41*
Mexican raids, 86
Mimbres, 133, *134*
Mimbres Archive, 129–130
Mimbres Foundation, 129–130
*Mimbres Painted Pottery* (Brody), 127
Mimbres pottery, *126*; animals on, *126*, *128*, *129*, 130, 130–133, *131*, *132*, *134*, *135*, P12; design choices, 130, 131–133; few depictions of corn, 130; humans on, 133–134, *135*, P12; kill holes, 128–129; Mimbres Archive, 129–130; origins, 127–128, *128*; rotational symmetry, 128; Style III, 127, *129*, 130, *131*, *132*, P12; uses, 128–129
Mimbres Pottery Image Digital Database (MimPIDD), 130
Mississippian culture, 52, 53
Mogollon culture, 128
Montgomery, Henry, 97
Monumental architecture, 52–53. *See also* Entries beginning with great houses
Moon, Maude, 8
Morning prayers (Zuni), 65, 69–70
Morss, Noel, 97, 101, 102
Motisinom, 72, 73, 74, 75–76
Mounds, 52, 110, 112, 115
Mountain sheep, *6*, 6–7
Moyle, Peter, 133
Mudheads, 67, 67–68
Mule deer, 24, *24*

Mummy Cave, P21
Museum of New Mexico, 43
Museum of Northern Arizona, 93

Nampeyo, *118*
National Park Service, 75, 93, 94–95
Navajo: cultural affiliation with Chaco remains, 75; death customs, 92–93; food, 85; historic nomadism, 89; oral tradition about Canyon de Chelley, 81–87; seasonal housing, 82; US government campaign against, 84–87, 89–90; World War II service, 94; in Wupatki region, 60, 92–95
Navajo Code Talkers, 94
Navajo Craftsman Exhibition, 93
Nawathis Village, 99
Neitzel, Jill, 49
New Alto, *52*
New fire ceremonies, 69
Nine Mile Canyon, 99
Nomadism: during Archaic, 4; distances covered during Paleoarchaic, 4; historic Navajo, 89; structured, 3
Non-Pueblo peoples near Mesa Verde region, 34
Nuva'ovi. *See* Wupatki Pueblo

Obsidian, 103
Old Pajarito Trail, 162
One-rod-and-bundle technique, 101
O'odham: dreaming as traveling, 149–150; history on calendar sticks, 155; oral history about Hohokam collapse, 116; spatial concepts, 149. *See also* Hohokam
O'odham songscapes: animal spirits and composition, 150; future of, 156–157; importance of, 156; instruments used, 154, *154*, *155*, *156*; Oriole song series, 150, 151–154, *152*, 156; relationships to journeys, 150–151
O'ohadag, 154–155, 156
Oral history: about Chaco Canyon, 73–74, *74*; about Wupatki Pueblo, 60; Hopi about Wupatki Pueblo, 59, 60, 61; Navajo about Ancestral Pueblo, 82–84; Navajo about Canyon de Chelley, 81–87; O'odham about Hohokam collapse, 116; O'odham calendar sticks as reminders, 155–156; warfare at Castle Rock Pueblo, 32
Oriole song series, 150, 151–154, *152*, 156
Oxidation pottery firing technique, 120

Paddle and anvil pottery technique, 119, *119*, 120
Pajarito Plateau: cavate pueblos, 37, 162; North Staircase, *165*; petroglyphs, 161, *161*; trails, 160, 160–164, **162**, *164*, P17
Pajarito Trails Project, 160
Paleoarchaic people, distances covered by, 4
Palisades, 27–28, *28*
Palmer, Edward, 97
Paquimé, 52
Parowan Valley site, 104
Patterson (shaman) bundle, 4
Peñasco Blanco, 48, 51
Peshlakai, Clyde, 93, 93–94, 95
Peshlakai Etsidi, *88*, 89, 90, *91*, 91–92, 93, *94*
Petroglyphs. *See* Rock art
"Pickleweed Winter, The" (Moon), 8
Pictographs. *See* Rock art
Pilling figurines, 101
Pit houses: ancestral Pueblo, 48; Fremont, 98–99, *100*, 105; Hohokam, 115; and population estimates, 48–49
Pit structures. *See* Kivas (architectural structures)
Place of the Snows. *See* Wupatki Pueblo
Plants: modern Mesa Verde, 20; native Mesa Verde food, 22–23,

23, 24; native Sonoran Desert food, *113*, 114, *114*; nondietary uses, 23–24; types gathered for food, 4–6, 8, 22–23, *23*, 24, 101, P10
Plaster: cavates, 37, 38, 39, 41; kivas, 39, 163; pit houses, 99
Platform mounds. *See* Mounds
Plazas, Wupatki Pueblo, 59–60
Polychrome pottery, 123–124. *See also* specific types
Ponderosa pine, 48
Population: Cahokia, 52; Chaco region, 48–49; clustering in Mesa Verde region, 28–29, 34, 58; decline due to violence, 34; density in Great Basin, 3; diversity at Wupatki, 60; and farming, 115; Hohokam diversity and size, 110–111, 115; and success of corn crops, 21–22
Pothunters, 58
Pot irrigation, 21
Pottery: along trails, 152; archaeological study of, 129; diverse styles at Wupatki, 60; food storage, 22; Fremont, 98, 101–102, P13; gender of producers, 133, 134; Hisat'sinom, *119*, 119–120, 121–123, P6; Hohokam, 111, 115, 124, 128, *128*; Hopi Nampeyo, *118*; Kayenta Puebloans, 121, 122–123, *123*, P6; predecessor, 4; preservation, 125; Pueblo Bonito, P9; Sierra Sin Agua, *119*, 119–125, *122*, P6; trade, 103; Zuni miniatures, 69. *See also* Mimbres pottery; specific types
Pottery of the Sierra Sin Agua, 119–125, P6; earliest, 119, 119–120, 121
Prayer meals, 68, 69
Prayer sticks, 68–69
Precipitation: Hohokam culture area, 114; Mesa Verde region, 21–22, 34; modern Mesa Verde, 20; San Juan Basin and Chaco Canyon, 11–12, *12*, 13–14, 15–16; Sonoran Desert, 115; Wupatki Pueblo, 55; Zuni rituals for, 67
Priesthoods, Zuni, 66, 67
Prince, LeBaron Bradford, 159, 165
Projectile points, P18
Pronghorn antelopes, *xii*, *1*, 1–3
Property beliefs, 3
Pueblo Alto, 51
Pueblo Bonito, P15; archaeological evidence of Hopi connections, 73, *74*; architecture, *44*, *47*, 47–48, *49*, *50*; construction time period, 45; as elite residence, 49; kivas, 78–79, P8; pottery, P9; and Wetherill, 45
Pueblo del Arroyo, 48, P15
Pueblo Revolt (1680-1692), 43
Pyramids, 52

Rainbow House (Frijoles Canyon), 163
Rain Priesthood, 66, 67
Rasping sticks, 154, *154*, *155*, 156
Red-on-brown pottery, 127
Red-on-Buff culture, 116
Red-on-buff pottery, 111, 115
Reduction pottery firing technique, 120, 124
Religion: art and rituals, 42; Chaco as built for ritual activities, 51; current Pueblo, 141–142; Hohokam ballcourts, 110, 111, 112, 115; Mimbres, 133, *134*; offerings, 68, 69, 78; and O'odham songscapes, 150, 156; repetitive rituals in, 163; representations of deities, 133; ritual fires to close buildings, 31; and rock art, 42, 137, 140, 141–142, 146, 147; sacred landscapes/sites, 156, 164; shamans, 1, 4, 147. *See also* Kivas (architectural structures); Zuni socio-religious system
Residences. *See* Housing
Rio Grande kivas, 48
Rituals. *See* Religion
Roads: Chaco, 51, 52. *See also* Trails
Roberts, Alexa, 89–95

Rock art: along trails, 152; ancestral Pueblo, 72, 76, 77, *139*, *140*, *142*, *143*, 144, 145–146; animals, 139, *139*, 140, 146, 147, P4; anthropomorphs, *136*, 137–138, *138*, *140*, 146, P22; Basketmaker, 72; Fremont, 102, 104–105, *105*; handprints, 138, *139*, 145, 146, *146*, 147; hunting depicted, 101; on interior walls, 39–40, *40*, *41*, 146, 147; Mesoamerican influence on Wupatki, 122; Navajo, 84; *o'ohadag* images, *154*; purposes, 131, 146–147, 161, *161*; and religion, 42, 137, 140, 141–142, 146, 147; of 1054 supernova, 76; textile designs, *143*, 145; warfare depicted, 31–33, *32*, 104
Roosevelt, Theodore, 92
Russell, Frank, 155

Sacred landscapes/sites, 156
Safety-in-numbers strategy, 28–29
Salapa. *See* Entries beginning with Mesa Verde
Salmon Ruin, 25, 147
Salt pilgrimages, 153
Salt River flooding, 116
Salzer, Matthew, 21
Sand Canyon Pueblo, P5; defensive measures, 28, *28*; food scarcity, 34; violence at, 27
Sand-dune dams, 12, *13*
San Francisco Mountain Gray Ware, *120*, 121, 125
San Juan Basin, 11–12, *12*
San Juan Red Ware, 123
Scalping, 33
Schaafsma, Polly, 104, 105, 137–147
School of American Research/School for Advanced Research, vii, 43
Scraping sticks, 154
Shalako, dance, 67–68; gods, 67; houses, 67, 68
Shamans, 1, 4, 147
Shell artifacts, 152–153
Shell jewelry, 60
Shoshone hunting, 1
Shrine of the Stone, 159
Sides (musician), *155*
Sierra Sin Agua people, 58, 60
Sierra Sin Agua pottery, Little Colorado White Ware, 124, *124*; Cohonina, *120*, 120–121; Flagstaff Black-on-white, *122*; Hisat'sinom, *119*, 119–120, 121–123, P6; Kayenta tradition, 123, *123*
Silentman, Irene, 81–87
Simms, Steve R., 1–10
Sings, 150
Sipapuni, 60
Skull flattening, 60
Sky Aerie Charnel House, 104
Sleeping Ute Mountain, *18*
Small-house sites: Chaco Canyon, 14, 15; Pueblo I, *46*
Snake Kiva, 39, 40, *41*
Snaketown, *111*, 117
Snead, James E., 159–165
Soalikee, Henry, *154*
Sofaer, Anna, 50, 51
Solstice Project, 51
Solstices and Zuni religious practices, 66–69, *67*
Song flowers, 154–155, *156*
Songscapes as trail maps, 149–150, 151–154, *152*, *153*
Sonoran Desert environment, *113*, *114*, 114–115
Southern Paiute, 4
Spatial concepts, 149
Spider Rock (Canyon de Chelley), 83
Split-twig figurines, P20
Stairs, *158*, 161, 163, 165

Index 173

Stein, John, 50, 51, 52
Steinaker Gap site, 98
Steward, Julian H., 97
Stockades, 27–28, *28*
Stone, Glenn, 60–61
Stone palettes, *112*
Storage: cavates, 40, 41, *42*; of food by modern Hopi, 74; food caches, 3–4; food jars, *22*; gourds for water, 85–86; granary rooms, *31*, 99, *99*, 100, 104; pits, 100, 115; rooms at great houses, 50, *50*; of tools, 3, *4*
Storm paths, 11–12, *12*
Summer solstice, 67
Sun Priest, 66
Sunset Brown Ware, 120
Sunset Crater, 58, 120, P25
Sunset Red Ware, 120
Supernova of 1054, 76, *76*
Swartz Ruin pottery, *126*, *135*
Swentzell, Rina, 133

Talus House (Frijoles Canyon), 37
Temperature: and corn-growing degree day units, 21; San Juan Basin and Chaco Canyon, 11–12, *12*; Sonoran Desert, 115
Territoriality, 3, *4*
*Tewusu. See* Zuni socio-religious system
Textiles: designs in rock art, *143*, 145; diverse styles at Wupatki, 60; Mesoamerican influence on, 122
Thompson, Mae, *81*, 81–87
Three Circle Red-on-white pottery, *127*
Tice grass, 23
Tobacco, 23
Tohono O'odham, 153
Tolchaco Mission, *90*, 91, *91*
Toll, H. Wolcott, 40
Tonto Basin, 111
Tools: hunting/weapons, 31, *32*, P18, P19; stones for, 2; storage, 3, *4*
Towers, 28
Trade: Chacoan, 73–74, 78; evidence along trails, 152–153; Fremont, 103–104; Hohokam-Mesoamerican, 113; Mimbres-Mesoamerican, 133; Sierra Sin Agua pottery imports, 123
Trail networks, 162
Trails, *153*, P17; ancestral Pueblo, *160*; categories, 162, 163–164; Chaco roads, 51, 52; construction, 160, 163–164; determining age, 161; during historic period, 162–163; major southwestern Native American, *150*; Pajarito Plateau, *160*, 162, **162**, 164, P17; petroglyphs as markers, 161, *161*; songscapes as maps of, 149–150, 151–154, *152*, *153*; stairs, *158*, 161, 163, *165*
Traps, animal, *xii*, *1*, 1–3, *6*, 6–7
*Tree Ring Dating of Archaeological Sites in the Chaco Canyon Region, New Mexico* (Bannister), 46
Trevathan, Wenda, 134
Trickster, 145
Tsankawi Pueblo: North Staircase, *165*; petroglyphs, 161, *161*; trails, *160*, 162, **162**, 164, P17
Tsegi Canyon rock art, *137*, *138*, *141*, *142*, *143*, *144*, *145*, *146*
Tsegi Orange Ware, 123–124
Turkeys, **33**, 34
Turner-Look site, 104
Turquoise, 68, *103*, 103–104
Tusayan Gray Ware, 123, *123*
Tusayan White Ware, *121*
Tuuwanasavi, 78

Tyuonyi, 38, 40

Una Vida, *47*, 47–48
Universal Trickster, 145

Van Dyke, Ruth, 50
Violence: causes, 33–35; at Cave 7, Whisker's Draw, 27, *32*; defensive measures in Mesa Verde region, 27–29, *28*, *29*, 30, 31, P14, P21; Fremont, 104; non-Pueblo attackers, 34; rock art as defensive measure, 147; rock art depictions, 31–33, *32*, 104; weapons, 31, *32*
Vivian, Gordon, 46, 47
Vivian, R. Gwinn, 11–17, 51

Warfare. *See* Violence
Water, P3; Chaco Canyon, *10*, 11–14, *12*, *13*; Hohokam resources, 115; modern Mesa Verde, 20; and population clustering, 29; Salt River flooding, 116; and seasonal residences, 82; spiritual aspects, 70–71; storage, 85–86. *See also* Irrigation
Water pockets, 11
Weapons, 31, *32*, P18
Weaving, 40, *42*
Wellman, Klaus, 144–145
Wetherill, Richard, 27, 45
Whisker's Draw, *26*, 27, *32*
White House Ruins (Canyon de Chelley), *80*
Wickiups, 2, *2*, 3–4, 99–100
Wilcox, David, 60
Windes, Tom, 13–14, 48, 51
Winter solstice, 67–69
Witchcraft, 84–85
Women: Hopi clan system, 78; and pottery, 133; in Zuni socio-religious system, 65, 66, 69
World War II, 94
Wukoki Pueblo, P7
Wupatki National Monument: enlarged, 93; established, 58, 89, 92; Little Box Canyon Pueblo, P1; Mesoamerican influence on rock art, *122*; Wukoki Pueblo, P7
Wupatki Pueblo, P23; amphitheater, *54*, 59; ballcourt, *54*, 59; construction, 58–59; departure from, 61–62; excavations, 61; Hopi names for, 58; as local form of Chaco great house, 60; oral traditions about, 59, 60, 61; plan, *56–57*; plazas, 59–60; precipitation, 55; as regional center, 60, 61; room block, *55*; Southwest at time of development, 58
Wupatki region Navajo, 92–95

Yellow Jacket, 48
Young, Jane, 146, 147
Yucca, 23
Yupköyvi, 73–74, *74*, 75–78

Zig-Zag Mountain, *148*, 149
Zuni: clay miniatures, 69; leadership, 66; oral tradition about Wupatki Pueblo, 60
Zuni River, 70–71
Zuni socio-religious system: components, 65–66; kivas, 65, 66; men's societies, 65, 66, 67, 68; personal conduct, 65, 66; rainfall rituals, 67; and rock art, 146; solstice rituals, 66–69, *67*; women's role, 65, 66; Zuni River as lifeline, 70–71